Progress in Computer Science and Applied Logic
Volume 1
Third Edition

Daniel H. Greene
Donald E. Knuth

Mathematics for the Analysis of Algorithms

Third Edition

1990

Birkhäuser
Boston • Basel • Berlin

Daniel H. Greene
Computer Science Laboratory
Xerox Palo Alto Research Center
Stanford, CA 94304, U.S.A.

Donald E. Knuth
Department of Computer Science
Stanford University
Stanford, CA 94305, U.S.A.

Library of Congress Cataloging-in-Publication Data
Greene, Daniel H., 1955-
Mathematics for the analysis of algorithms / Daniel H. Greene,
Donald E. Knuth. – 3rd ed.
p. cm. – (Progress in computer science and applied logic ; no. 1)
Includes bibliographical references and index.
ISBN 0-8176-3515-7 (alk. paper)
1. Electronic digital computers – Programming. 2. Algorithms.
I. Knuth, Donald Ervin, 1938- . II. Title. III. Series.
QA76.6..G7423 1990
005.1 – dc20 90-517

Printed on acid-free paper.

ISBN 0-8176-3515-7
ISBN 3-7643-3515-7

Photocomposed copy prepared with TEX using the METAFONT system.
Printed and bound by R.R. Donnelly & Sons Harrisonburg, VA, U.S.A.
Printed in the U.S.A.

9 8 7 6 5 4 3 2 1

Preface

This monograph is derived from an advanced course in computer science at Stanford University on the analysis of algorithms. The course presents examples of the major paradigms used in the precise analysis of algorithms, emphasizing some of the more difficult techniques. Much of the material is drawn from the starred sections of *The Art of Computer Programming*, Volume 3 [Knuth III].

Analysis of algorithms, as a discipline, relies heavily on both computer science and mathematics. This report is a mathematical look at the synthesis—emphasizing the mathematical perspective, but using motivation and examples from computer science. It covers binomial identities, recurrence relations, operator methods and asymptotic analysis, hopefully in a format that is terse enough for easy reference and yet detailed enough to be of use to those who have not attended the lectures. However, it is assumed that the reader is familiar with the fundamentals of complex variable theory and combinatorial analysis.

Winter 1980 was the fourth offering of Analysis of Algorithms, and credit is due to the previous teachers and staff—Leo Guibas, Scott Drysdale, Sam Bent, Andy Yao, and Phyllis Winkler—for their detailed contributions to the documentation of the course. Portions of earlier handouts are incorporated in this monograph. Harry Mairson, Andrei Broder, Ken Clarkson, and Jeff Vitter contributed helpful comments and corrections, and the preparation of these notes was also aided by the facilities of Xerox corporation and the support of NSF and Hertz graduate fellowships.

In this third edition we have made a few improvements to the exposition and fixed a variety of minor errors. We have also added several new appendices containing exam problems from 1982 and 1988.

—D.H.G. and D.E.K.

Contents

Chapter 1

Binomial Identities

1.1 Summary of Useful Identities

So that the identities themselves do not become buried on an obscure page, we summarize them immediately:

$$(x+y)^n = \sum_k \binom{n}{k} x^k y^{n-k}, \qquad \begin{array}{c}\text{integer } n \\ \text{or } n \text{ real and } |x/y| < 1\end{array} \tag{1.1}$$

$$\binom{r}{k} = \binom{r-1}{k} + \binom{r-1}{k-1}, \qquad \begin{array}{c}\text{real } r \\ \text{integer } k\end{array} \tag{1.2}$$

$$\binom{n}{k} = \binom{n}{n-k}, \qquad \begin{array}{c}\text{integer } n \ge 0 \\ \text{integer } k\end{array} \tag{1.3}$$

$$\binom{r}{k} = \frac{r}{k}\binom{r-1}{k-1}, \qquad \begin{array}{c}\text{real } r \\ \text{integer } k \ne 0\end{array} \tag{1.4}$$

$$\sum_{k=0}^{n} \binom{r+k}{k} = \binom{r+n+1}{n}, \qquad \begin{array}{c}\text{real } r \\ \text{integer } n \ge 0\end{array} \tag{1.5}$$

$$\sum_{k=0}^{n} \binom{k}{m} = \binom{n+1}{m+1}, \qquad \text{integer } m, n \ge 0 \tag{1.6}$$

$$\binom{-r}{k} = (-1)^k \binom{r+k-1}{k}, \qquad \begin{array}{c}\text{real } r \\ \text{integer } k\end{array} \tag{1.7}$$

$$\binom{r}{m}\binom{m}{k} = \binom{r}{k}\binom{r-k}{m-k}, \qquad \begin{array}{c}\text{real } r\\ \text{integer } m, k\end{array} \tag{1.8}$$

$$\sum_k \binom{r}{k}\binom{s}{n-k} = \binom{r+s}{n}, \qquad \begin{array}{c}\text{real } r, s\\ \text{integer } n\end{array} \tag{1.9}$$

$$\sum_k \binom{r}{k}\binom{s}{n+k} = \binom{r+s}{r+n}, \qquad \begin{array}{c}\text{integer } n, \text{ real } s\\ \text{integer } r \geq 0\end{array} \tag{1.10}$$

$$\sum_k \binom{r}{k}\binom{s+k}{n}(-1)^k = (-1)^r\binom{s}{n-r}, \qquad \begin{array}{c}\text{integer } n, \text{ real } s\\ \text{integer } r \geq 0\end{array} \tag{1.11}$$

$$\sum_{k=0}^{r} \binom{r-k}{m}\binom{s+k}{n} = \binom{r+s+1}{m+n+1}, \qquad \begin{array}{c}\text{integer } m, n, r, s \geq 0\\ n \geq s\end{array} \tag{1.12}$$

Parameters called real here may also be complex.

One particularly confusing aspect of binomial coefficients is the ease with which a familiar formula can be rendered unrecognizable by a few transformations. Because of this chameleon character there is no substitute for practice of manipulations with binomial coefficients. The reader is referred to Sections 5.1 and 5.2 of [GKP] for an explanation of the formulas above and for examples of typical transformation strategy.

1.2 Deriving the Identities

Here is an easy way to understand many of the identities that do not include an alternating -1. The number of monotonic paths through a rectangular lattice with sides m and n is $\binom{m+n}{m}$. By cutting the lattice along different axes, and counting the paths according to where they cross the cut, the identities are derived. The pictures below show different ways of partitioning the paths and the parameter k used in the sum.

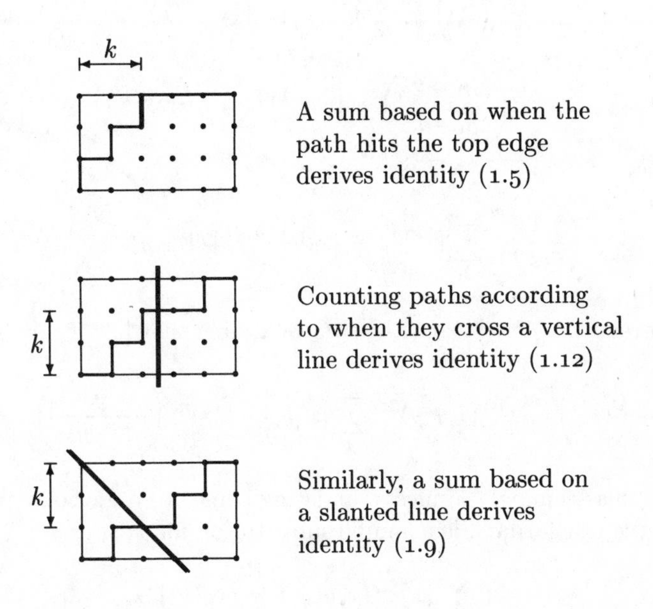

A sum based on when the path hits the top edge derives identity (1.5)

Counting paths according to when they cross a vertical line derives identity (1.12)

Similarly, a sum based on a slanted line derives identity (1.9)

More complicated identities can be derived by successive applications of the identities given on pages 1 and 2. One example appears in "A trivial algorithm whose analysis isn't," by A. Jonassen and D. E. Knuth [Jonassen 78], where the sum

$$S = \sum_k \binom{m}{k} \left(-\frac{1}{2}\right)^k \binom{2k}{k} \tag{1.13}$$

is evaluated by a lengthy series of elementary transformations. Instead of repeating that derivation, let us consider instead a derivation suggested by I. Gessel. He attributes this elegant technique, the "method of coefficients," to G. P. Egorychev.

First replace k by $m-k$, giving

$$S = \sum_k \binom{m}{k} \left(-\frac{1}{2}\right)^{m-k} \binom{2m-2k}{m-k}. \tag{1.14}$$

Using the notation $[x^n]\, f(x)$ for the coefficient of x^n in $f(x)$, we can express portions of the sum with generating functions:

$$\binom{m}{k} \left(-\frac{1}{2}\right)^{-k} = [x^k]\,(1-2x)^m \tag{1.15}$$

$$\binom{2m-2k}{m-k} = [y^{m-k}]\,(1+y)^{2m-2k}. \tag{1.16}$$

The whole sum is

$$S = \left(-\frac{1}{2}\right)^m \sum_k [x^k]\,(1-2x)^m [y^{m-k}]\,(1+y)^{2m-2k}. \tag{1.17}$$

We can remove $[y^{m-k}]$ from the sum by noting that $[y^{m-k}] = [y^m]\, y^k$:

$$S = \left(-\frac{1}{2}\right)^m [y^m]\,(1+y)^{2m} \sum_k [x^k]\,(1-2x)^m \left(\frac{y}{(1+y)^2}\right)^k. \tag{1.18}$$

Finally, this seemingly aimless wandering comes to a glorious finish. The sum in the last formula is a simple substitution for x, since

$$\sum_k [x^k]\, f(x) g(y)^k = f(g(y)) \tag{1.19}$$

when f is analytic. The solution follows immediately:

$$S = (-2)^{-m} [y^m]\,(1+y)^{2m} \left(1 - \frac{2y}{(1+y)^2}\right)^m = (-2)^{-m} [y^m]\,(1+y^2)^m; \tag{1.20}$$

$$S = \begin{cases} 2^{-m}\binom{m}{m/2}, & m \text{ even;} \\ 0, & m \text{ odd.} \end{cases} \tag{1.21}$$

A simpler approach to this problem has been pointed out by C. C. Rousseau, who observes that $\binom{2k}{k}$ is the coefficient of x^0 in $(x+x^{-1})^{2k}$, hence S is the coefficient of x^0 in $\left(1-(x+x^{-1})^2/2\right)^m$.

From a theoretical standpoint, it would be nice to unify such identities in one coherent scheme, much as the physicist seeks a unified field theory. No single scheme covers everything, but there are several "meta" concepts that explain the existence of large classes of binomial identities. We will briefly describe three of these: inverse relations, operator calculus, and hypergeometric series.

1.3 Inverse Relations

One of the simplest set of inverse relations is the pair

$$a_n = \sum_k (-1)^k \binom{n}{k} b_k, \qquad b_n = \sum_k (-1)^k \binom{n}{k} a_k, \tag{1.22}$$

which follows from the orthogonal relation

$$[n = k] = \sum_{j=0}^{n} (-1)^{j+k} \binom{n}{j} \binom{j}{k}. \tag{1.23}$$

(A logical relation in brackets evaluates to 1 if true, 0 if false. We assume that n is a nonnegative integer.) This formula is just a specialization of equation (1.11) with s equal to zero. In general an inverse relation will pair two series so that individual terms of one can be computed from the terms of the other. There will always be an associated orthogonal relation.

In his book *Combinatorial Identities*, John Riordan devotes several chapters to inverse relations. Since inverse relations are even more likely to change appearance than the binomial identities we have seen already, care must be taken to recognize relations that are basically the same. For this purpose Riordan describes several transformations and then groups equivalent inverse pairs into broad categories. His transformations and classifications are summarized below.

Since we are working with a pair of equations, we can move terms from one equation to another by replacements like $b'_k = (-1)^k b_k$, obtaining a new pair

$$a_n = \sum_k \binom{n}{k} b'_k, \qquad b'_n = \sum_k (-1)^{k+n} \binom{n}{k} a_k. \tag{1.24}$$

An inverse relation corresponds to a pair of lower triangular matrices whose product is the identity. By reflecting across the diagonal we can derive yet another pair

$$a_n = \sum_{k \geq n} \binom{k}{n} b_k, \qquad b_n = \sum_{k \geq n} (-1)^{k+n} \binom{k}{n} a_k. \tag{1.25}$$

Finally, note that we can multiply both sides of the orthogonal relation (1.23) by almost any function that is unity when $n = k$, without affecting the orthogonal character of the equation.

The last equation, (1.25), has an extremely useful combinatorial significance. Suppose we have a large collection of random events. Let b_n be the probability that *exactly* n events occur, and let a_n be the sum of the probability of n simultaneous events taken over all selections of n events. Roughly speaking a_n can be viewed as a sloppy way of computing the probability that exactly n events occur since it makes no allowance for the possibility of more than n events. The left side of (1.25) shows how a_n is inflated. However, a_n is often easier to compute and the right hand side of equation (1.25), the "principle of inclusion and exclusion," provides a practical way of obtaining b_n.

Equations (1.22), (1.24) and (1.25) belong to the simplest class of inverse relations. [Riordan 68] lists several other classes like the Chebyshev type:

$$a_n = \sum_{k=0}^{\lfloor n/2 \rfloor} \binom{n}{k} b_{n-2k}, \qquad b_n = \sum_{k=0}^{\lfloor n/2 \rfloor} (-1)^k \frac{n}{n-k} \binom{n-k}{k} a_{n-2k}. \tag{1.26}$$

Not surprisingly, these inverse relations are often associated with their namesakes among the orthogonal polynomials used in interpolation.

The Gould class of inverse relations,

$$f_n = \sum_k (-1)^k \binom{n}{k} \binom{a+bk}{n} g_k, \tag{1.27}$$

$$g_n \binom{a+bn}{n} = \sum_k (-1)^k \frac{a+bk-k}{a+bn-k} \binom{a+bn-k}{n-k} f_k, \tag{1.28}$$

has a very curious property. A Chinese mathematician L. Hsu recently discovered that the binomial coefficients containing a and b are inessential to the functioning of the inversion. In fact if we choose $\{a_i\}$ and $\{b_i\}$ to be any two sequences of numbers such that

$$\psi(x,n) = \prod_{i=1}^{n} (a_i + b_i x) \neq 0, \qquad \text{integer } x, n \geq 0, \tag{1.29}$$

we obtain a general inversion:

$$f_n = \sum_k (-1)^k \binom{n}{k} \psi(k,n)\, g_k, \tag{1.30}$$

$$g_n = \sum_k (-1)^k \binom{n}{k} (a_{k+1} + k\, b_{k+1}) \psi(n, k+1)^{-1} f_k. \tag{1.31}$$

Another well known pair of inverse relations uses Stirling numbers:

$$a_n = \sum_{k=0}^{n} (-1)^{n-k} \left[{n \atop k}\right] b_k, \quad \left[{n \atop k}\right] \equiv \text{Stirling numbers of the first kind;} \tag{1.32}$$

$$b_n = \sum_{k=0}^{n} \left\{{n \atop k}\right\} a_k, \quad \left\{{n \atop k}\right\} \equiv \text{Stirling numbers of the second kind.} \tag{1.33}$$

Here a_n is usually $x^{\underline{n}}$ and b_n is x^n, so that these formulas convert between factorial powers and ordinary powers of x.

We cannot explore all the inverse relations here, but it is worth noting that many generating functions can be converted to inverse relations. A pair of power series $z(x)$ and $z^*(x)$ such that $z(x)\, z^*(x) = 1$ provides a pair of relations:

$$a(x) = z(x)\, b(x), \quad \text{and} \quad b(x) = z^*(x)\, a(x). \tag{1.34}$$

For example, we can let $z(x) = (1-x)^{-p}$ and $z^*(x) = (1-x)^p$; clearly $z(x)\, z^*(x) = 1$, so we can proceed to compute formulas for the coefficients in $a(x)$ and $b(x)$:

$$a_n = \sum_k (-1)^k \binom{-p}{k} b_{n-k}, \qquad b_n = \sum_k (-1)^k \binom{p}{k} a_{n-k}. \tag{1.35}$$

This pair is a member of the Gould class of inverse relations.

Inverse relations are partially responsible for the proliferation of binomial identities. If one member of an inverse pair can be embedded in a binomial identity, then the other member of the pair will often provide a new identity. Inverse relations can also enter directly into the analysis of an algorithm. The study of radix exchange sort, for example, uses the simple set of relations (1.22) introduced at the beginning of this section. For details see [Knuth III; exercises 5.2.2–36 and 5.2.2–38].

1.4 Operator Calculus

There is a striking similarity between the integral

$$\int_a^b x^n \, dx = \left. \frac{x^{n+1}}{n+1} \right|_a^b \tag{1.36}$$

and the sum

$$\sum_{a \le x < b} x^{\underline{n}} = \left. \frac{x^{\underline{n+1}}}{n+1} \right|_a^b , \tag{1.37}$$

where the underlined superscript, $x^{\underline{n}} = x(x-1)(x-2)\ldots(x-n+1)$, denotes a falling factorial. The latter sum is merely a variation of equation (1.6) in a form that is easy to remember. It is certainly easier to remember than the formula for sums of powers found on page 47.

The similarity of equations (1.36) and (1.37) is a consequence of the facts that $Dx^n = nx^{n-1}$ and $\Delta x^{\underline{n}} = nx^{\underline{n-1}}$, where D and Δ are the operators of differentiation and difference that are inverse to $\int$ and $\sum$: $D\,p(x) = p'(x)$ and $\Delta\,p(x) = p(x+1) - p(x)$. We can extend such analogies much further; Rota, for example, gives the following generalization of Taylor's theorem:

Definitions. Let E^a be the shift operator, $E^a p(x) = p(x+a)$. An operator Q is a delta operator if it is shift invariant ($Q\,E^a = E^a\,Q$) and if Qx is a nonzero constant. Such an operator has a sequence of basic polynomials defined as follows:

i) $p_0(x) = 1$
ii) $p_n(0) = 0, \quad n > 0$
iii) $Q\,p_n(x) = n\,p_{n-1}(x)$.

The third property means that whenever Q is applied to its basic polynomials the result is similar to D applied to $1, x, x^2, \ldots$. For example, Δ is a delta operator with basic polynomials $x^{\underline{n}} = x(x-1)\ldots(x-n+1)$.

Taylor's Theorem.

$$T = \sum_k \frac{a_k}{k!} Q^k \tag{1.38}$$

where

T is any shift invariant operator;

Q is any delta operator with basic polynomials $p_k(x)$;

$a_k = T\,p_k(x)\big|_{x=0}$.

When $T = E^a$ and $Q = D$, this reduces to the well known Taylor formula. By changing Q to Δ, the difference operator, we obtain Newton's expansion of $T = E^a$,

$$p(x+a) = \sum_k \frac{a^{\underline{k}}}{k!} \Delta^k p(x). \qquad (1.39)$$

Newton's expansion is a useful tool for proving binomial identities. Equation (1.9), for example, is an expansion of $p(s+r) = (s+r)^{\underline{m}}$.

A full exposition of operator calculus and its relation to binomial identities can be found in [Rota 75]. The reader will also notice the close relationship between discrete and continuous analysis in Chapter 2, where difference equations resemble differential equations, and in Section 4.2 on Stieltjes integration, where floor and ceiling functions are "integrated" to produce sums.

1.5 Hypergeometric Series

The geometric series $1 + z + z^2 + \cdots = 1/(1-z)$ can be generalized to a hypergeometric series

$$F(a,b;c;z) = 1 + \frac{ab}{c}\frac{z}{1!} + \frac{a(a+1)\,b(b+1)}{c(c+1)}\frac{z^2}{2!} + \frac{a^{\overline{n}} b^{\overline{n}}}{c^{\overline{n}}}\frac{z^n}{n!} + \cdots, \qquad (1.40)$$

where the overlined superscript $a^{\overline{n}} = a(a+1)(a+2)\ldots(a+n-1)$ signifies a rising factorial power. The semicolons in the parameter list of F indicate that there are two numerator parameters (a,b) and one denominator parameter (c). The hypergeometric series in this example can be further generalized to an arbitrary number of numerator and denominator parameters.

The standardization afforded by hypergeometric series has shed much light on the theory of binomial identities. For example, identities (1.5), (1.10) and (1.11) are all consequences of Vandermonde's theorem:

$$F(a,-n;c;1) = \frac{(c-a)^{\overline{n}}}{c^{\overline{n}}} \qquad \text{integer } n > 0. \qquad (1.41)$$

The negative integer argument $-n$ terminates the otherwise infinite series, allowing us to express (1.10) as a variation of this formula:

$$\frac{s^{\underline{n}}}{n!} F(-r, -s+n; n+1; 1) = \frac{s^{\underline{n}}}{n!}\frac{(s+1)^{\overline{r}}}{(n+1)^{\overline{r}}} = \binom{r+s}{r+n}. \qquad (1.42)$$

More information on hypergeometric series can be found in [Bailey 35], [Henrici I], and [GKP].

1.6 Identities with the Harmonic Numbers

Harmonic numbers occur frequently in the analysis of algorithms and there are some curious identities that involve both binomial coefficients and harmonic numbers. The commonly used identities are summarized here.

$$H_n = \sum_{k=1}^{n} \frac{1}{k} \tag{1.43}$$

$$\sum_{k=1}^{n} H_k = (n+1)H_n - n \tag{1.44}$$

$$\sum_{k=1}^{n} \binom{k}{m} H_k = \binom{n+1}{m+1} \left(H_{n+1} - \frac{1}{m+1} \right) \tag{1.45}$$

$$\sum_{k=1}^{n} \binom{n}{k} x^k H_k = (x+1)^n \left(H_n - \ln\left(1 + \frac{1}{x}\right)\right) + \epsilon, \qquad \begin{array}{c} x > 0 \\ 0 < \epsilon < \frac{1}{x(n+1)} \end{array} \tag{1.46}$$

$$\frac{1}{(1-z)^{m+1}} \ln\left(\frac{1}{1-z}\right) = \sum_{n \ge 0} (H_{n+m} - H_m) \binom{n+m}{n} z^n \tag{1.47}$$

$$\frac{1}{(1-z)^{m+1}} \ln\left(\frac{1}{1-z}\right)^2 = \sum_{n \ge 0} \Big((H_{n+m} - H_m)^2 - (H_{n+m}^{(2)} - H_m^{(2)}) \Big) \binom{n+m}{n} z^n \tag{1.48}$$

The last two identities, along with a generalization to higher powers, appear in [Zave 76]. We can regard them as identities valid for complex values of m, with $H_{n+m} - H_m = \frac{1}{m+1} + \frac{1}{m+2} + \cdots + \frac{1}{m+n}$; see the solution of problem 2(g), midterm exam II, on pages 105–106 below.

Chapter 2

Recurrence Relations

2.1 Linear Recurrence Relations

Recurrence relations are traditionally divided into two classes: A recurrence with "finite history" depends on a fixed number of earlier values,

$$x_n = f(x_{n-1}, x_{n-2}, \ldots, x_{n-m}), \qquad n \geq m. \tag{2.1}$$

An equation that depends on all preceding values has a "full history."

The simplest recurrences have a finite history, and f is a linear function with constant coefficients. Here the terminology parallels differential equation theory; we distinguish between the "homogeneous" and the "nonhomogeneous" situations depending on the presence of an extra term $g(n)$:

$$c_0 x_n + c_1 x_{n-1} + \cdots + c_m x_{n-m} = g(n). \tag{2.2}$$

There are two classic treatises on the calculus of finite differences, one by Jordan [Jordan 60] and the other by Milne-Thomson [Mil-Thom 33]. Although the emphasis of these early works was on approximation and solution of differential equations—problems in the mainstream of numerical analysis rather than analysis of algorithms—much can be learned from this theory. We recommend a recent summary by Spiegel [Spiegel 71] and *An Introduction to Computational Combinatorics* by Page and Wilson [Page 79].

Within this section references are given to additional examples of the solution of recurrence relations from [Knuth I] and [Knuth III]. The last part of the section, on the repertoire approach to full history equations, was introduced in a paper by D. Knuth and A. Schönhage [Knuth 78].

2.1.1 Finite History

2.1.1.1 Constant Coefficients

The constant coefficient problem is a beautiful example of the use of generating functions to solve recurrence relations. Rather than attempting to find x_n directly, we construct a function $G(z)$ with coefficients x_n in its power series expansion:

$$G(z) = \sum_k x_k\, z^k. \tag{2.3}$$

The recurrence relation is converted to an equation in $G(z)$ and solved by whatever means are applicable. This is best explained with an example,

$$x_{n+2} - 3x_{n+1} + 2x_n = n, \qquad x_0 = x_1 = 1. \tag{2.4}$$

First we multiply by z^{n+2} and sum over all n, obtaining

$$\sum_{n\geq 0} x_{n+2} z^{n+2} - 3z \sum_{n\geq 0} x_{n+1} z^{n+1} + 2z^2 \sum_{n\geq 0} x_n z^n = \sum_{n\geq 0} n z^{n+2}. \tag{2.5}$$

The first sum is $G(z)$ missing its first two terms. The next two sums are similarly close to $G(z)$, and the right side of the equation can be expressed in closed form as $z^3/(1-z)^2$. (This follows from the binomial theorem, equation (1.1), when $(x+y)^n = (1-z)^{-2}$. A list of standard closed forms for generating functions appears in [GKP; Chapter 7].)

Putting everything together in one formula for $G(z)$ gives

$$G(z) - z - 1 - 3z\,(G(z) - 1) + 2z^2 G(z) = \frac{z^3}{(1-z)^2}. \tag{2.6}$$

And this is easy to solve for $G(z)$:

$$G(z) = \frac{z^3}{(1-z)^2(1-3z+2z^2)} + \frac{-2z+1}{(1-3z+2z^2)}. \tag{2.7}$$

We would like to recover the coefficient of z^n in $G(z)$. If the denominators of the fractions in $G(z)$ were linear, the recovery problem would be simple: each term would be a geometric series. This is not the case in the example we have, but by expressing our solution for $G(z)$ in partial fractions we obtain a manageable form:

$$G(z) = \frac{1}{1-2z} + \frac{1}{(1-z)^2} - \frac{1}{(1-z)^3}. \tag{2.8}$$

Note that the only nonlinear denominators are higher powers of a linear factor. These terms can be expanded with the binomial theorem, and x_n is easily computed:

$$x_n = 2^n - \frac{n^2+n}{2}. \tag{2.9}$$

Partial fractions are powerful enough to deal with all linear recurrences with constant coefficients. For simplicity, however, we will discuss a different approach found in [Spiegel 71] and many of the older references. The approach is based on trial solutions and is similar to the solution of differential equations. In certain instances this second approach will provide fast answers, but the rules often seem like black magic, and the puzzled reader will have to return to the underlying partial fraction theory to understand why these "rules of thumb" work.

A) Homogeneous Equations.

$$c_0x_n + c_1x_{n-1} + \cdots + c_mx_{n-m} = 0, \qquad n \geq m. \tag{2.10}$$

We try $x_n = r^n$, and obtain an mth degree polynomial in r. Let $r_1, \ldots, r_m$ be the roots of this polynomial. The "general solution" is

$$x_n = k_1r_1^n + k_2r_2^n + \cdots + k_mr_m^n, \tag{2.11}$$

where the k_i are constants determined by the initial values.

Multiple roots are accommodated by prefacing the terms of the general solution with powers of n. Suppose that $r_1 = r_2 = r_3$; then the adjusted solution would be

$$x_n = k_1r_1^n + k_2n\,r_1^n + k_3n^2r_1^n. \tag{2.12}$$

B) Nonhomogeneous Equations.

$$c_0x_n + c_1x_{n-1} + \cdots + c_mx_{n-m} = g(n). \tag{2.13}$$

First remove $g(n)$ and obtain the general solution to the homogeneous equation. Add to this homogeneous solution any "particular" solution to the nonhomogeneous equation.

A particular solution can be found by the method of "undetermined coefficients." The idea is to use a trial solution with unspecified coefficients and then solve for these coefficients. The nature of the trial solution depends on the form of $g(n)$:

Form of $g(n)$:	Trial Solution:
α^n	$k\,\alpha^n$ (multiply by n if α is a root)
$p(n)$	polynomial of the same degree

2.1.1.2 Variable Coefficients

There is no guaranteed solution to the variable coefficient problem, but there are several methods worth trying:

A) Summation Factors.

If the equation is "first order,"

$$a(n)x_n = b(n)\,x_{n-1} + c(n), \qquad n \geq 1, \tag{2.14}$$

then it can be reduced to a summation. First multiply both sides by the summation factor

$$F(n) = \frac{\prod_{i=1}^{n-1} a(i)}{\prod_{j=1}^{n} b(j)}. \tag{2.15}$$

Then the recurrence becomes

$$y_n = y_{n-1} + F(n)\,c(n), \tag{2.16}$$

where $y_n = b(n+1)\,F(n+1)\,x_n$. The last recurrence allows us to express x_n as a sum:

$$x_n = \frac{x_0 + \sum_{i=1}^{n} F(i)\,c(i)}{b(n+1)\,F(n+1)}. \tag{2.17}$$

See [Knuth III; page 121] and [Lueker 80] for illustrations of this technique.

B) Generating Functions.

Variable coefficients are amenable to a generating function attack. If the coefficients are polynomials, the generating function is differentiated to obtain the desired variability. Let us look at a relatively simple problem to get a feeling for what is involved:

$$(n+1)x_{n+1} - (n+r)x_n = 0. \tag{2.18}$$

Multiplying by z^n and summing over all n will bring us closer to a formula in $G(z)$:

$$\sum_n (n+1)x_{n+1}\, z^n - \sum_n (n+r)x_n\, z^n = 0. \tag{2.19}$$

Using the derivative of $G(z)$ and multiplication by z for shifting, we obtain a differential equation,

$$(1-z)\, G'(z) - r\, G(z) = 0. \tag{2.20}$$

In general any recurrence with coefficients that are polynomial in n can be converted to a differential equation like (2.20). In this case, the coefficients of the solution, $G(z) = (1-z)^{-r}$, can be recovered by the binomial theorem:

$$x_n = (-1)^n \binom{-r}{n} = \binom{r-1+n}{n}. \tag{2.21}$$

More difficult problems will present the analyst with a wide variety of equations in $G(z)$. While these are not always differential equations, the reader is referred to [Boyce 69] for those differential equations that do occur.

C) Reduction of Order.

If we are fortunate enough to factor the difference equation, then we can attempt to solve each factor separately. For example, the difference equation

$$y_{k+2} - (k+2)\, y_{k+1} + k\, y_k = k \tag{2.22}$$

can be written in operator notation:

$$\big(E^2 - (k+2)E + k\big) y_k = k. \tag{2.23}$$

And the operator can be factored so that

$$(E-1)(E-k)y_k = k. \tag{2.24}$$

If we first solve the equation

$$(E-1)z_k = k, \tag{2.25}$$

which has the simple answer $z_k = \binom{k}{2}$, then we have reduced the order, leaving a first order equation:

$$(E-k)\, y_k = \binom{k}{2}. \tag{2.26}$$

Using $F(n) = 1/n!$ as a summing factor, the last equation can be solved:

$$y_n = \frac{(n-1)!}{2} \sum_{k=1}^{n-3} \frac{1}{k!}. \tag{2.27}$$

For simplicity we will omit the discussion of initial conditions; see [Spiegel 71; page 176] for a solution of this example with initial conditions $y_1 = 0$ and $y_2 = 1$.

All three approaches to the variable coefficient problem have serious shortcomings. The summation factor may yield an inscrutable sum, and the generating function can produce an equally intractable differential equation. And alas, there is no certain way to factor an operator equation to apply the reduction of order technique. The variable coefficient equation is a formidable problem; we will have to return to it later in the exploration of asymptotic approximations.

2.1.2 Full History

2.1.2.1 Differencing

The differencing strategy eliminates full history by subtracting suitable combinations of adjacent formulas. For example, [Knuth III; page 120] solves the equation

$$x_n = f_n + \frac{2}{n} \sum_{k=0}^{n-1} x_k \tag{2.28}$$

by subtracting

$$n\, x_n = n\, f_n + 2 \sum_{k=0}^{n-1} x_k \tag{2.29}$$

from

$$(n+1)x_{n+1} = (n+1)f_{n+1} + 2 \sum_{k=0}^{n} x_k, \tag{2.30}$$

yielding a first order variable coefficient problem. Note how the two formulas have been carefully rearranged to eliminate the sum. In complex situations, several differences may be necessary to remove the history. See, for example, [Knuth III; exercise 6.2.2–7].

2.1.2.2 By Repertoire

In the next approach we take advantage of the linearity of the recurrence and construct the desired solution from a repertoire of simple solutions. Several recurrences in the analysis of algorithms have the form

$$x_n = a_n + \sum_{0 \le k \le n} p_{nk}(x_k + x_{n-k}), \qquad \sum_k p_{nk} = 1. \tag{2.31}$$

If we also know that

$$y_n = b_n + \sum_{0 \le k \le n} p_{nk}(y_k + y_{n-k}), \tag{2.32}$$

then by linearity an equation with additive term $\alpha\, a_n + \beta\, b_n$ will have the solution $\alpha\, x_n + \beta\, y_n$.

The crucial idea is this: We choose x_n first so as to make the sum tractable, *then* we see what additive term a_n results from the x_n. This is exactly backwards from the original problem, where a_n is given and x_n

is sought. However, once we have built up a repertoire of enough additive terms, the original a_n can be constructed by linear combination.

For example, consider the recurrence associated with median-of-three quicksort:

$$x_n = n + 1 + \sum_{1 \le k \le n} \frac{\binom{k-1}{1}\binom{n-k}{1}}{\binom{n}{3}} (x_{k-1} + x_{n-k}). \qquad (2.33)$$

The ordinary quicksort algorithm is modified so that three elements are chosen and the median of these three is used for the partitioning phase of the algorithm. In ordinary quicksort each partition size is equally likely. This modification makes it slightly more likely that the partition will split into similar sized parts, because $p_{nk} = \binom{k-1}{1}\binom{n-k}{1}/\binom{n}{3}$ is largest when k is near $n/2$.

At first we observe that the sum is symmetric and we replace the additive term, $n+1$, by a_n in preparation for the repertoire approach:

$$x_n = a_n + \frac{2}{\binom{n}{3}} \sum_{1<k<n} (k-1)(n-k)x_{k-1}. \qquad (2.34)$$

Choosing x_n equal to the falling factorial $(n-1)^{\underline{s}}$ makes the sum in equation (2.34) easy to compute:

$$\begin{aligned}
(n-1)^{\underline{s}} &= a_n + \frac{12}{n^{\underline{3}}} \sum_{1<k<n} (n-k)(k-1)^{\underline{s+1}} \\
&= a_n + \frac{12(s+1)!}{n^{\underline{3}}} \sum_{1<k<n} \binom{n-k}{1}\binom{k-1}{s+1} \\
&= a_n + \frac{12(s+1)!}{n^{\underline{3}}} \binom{n}{s+3}; \qquad (2.35)\\
a_n &= (n-1)^{\underline{s}} - \frac{12}{(s+2)(s+3)}(n-3)^{\underline{s}}. \qquad (2.36)
\end{aligned}$$

Now we have a family of solutions parameterized by s.

	x_n	a_n
$s=0$	1	-1
$s=1$	$(n-1)$	2
$s=2$	$(n-1)(n-2)$	$\frac{2n^2+6n-26}{5}$

However, the family is inadequate; it lacks a member with linear a_n. The possibilities for a_n jump from constant to $\Theta(n^2)$ and unfortunately the a_n that we wish to reconstruct is $\Theta(n)$. On reflection, this is not surprising. We expect the solution of this divide and conquer style of iteration to be $O(n \log n)$ and yet we have limited the possibilities for x_n to polynomials in n. So to expand our family of solutions we introduce the harmonic numbers, H_n, which are also easy to sum and will contribute $O(\log n)$ factors to the solutions. The new family is computed using $x_n = (n-1)^{\underline{t}} H_n$ in equation (2.34) and solving for a_n.

$$\begin{aligned}(n-1)^{\underline{t}} H_n &= a_n + \frac{12}{n^{\underline{3}}} \sum_{1<k<n} (n-k)(k-1)^{\underline{t+1}} H_{k-1} \\ &= a_n + \frac{12}{n^{\underline{3}}} \left(\sum_{1<k<n} n(k-1)^{\underline{t+1}} H_{k-1} \right. \\ &\qquad \left. - \sum_{1<k<n} k^{\underline{t+2}} H_k + \sum_{1<k<n} (k-1)^{\underline{t+1}} \right) \\ &= a_n + \frac{12}{n^{\underline{3}}} \left(\frac{n^{\underline{t+3}}}{t+2} \left(H_{n-1} - \frac{1}{t+2} \right) \right. \\ &\qquad \left. - \frac{n^{\underline{t+3}}}{t+3} \left(H_n - \frac{1}{t+3} \right) + \frac{(n-1)^{\underline{t+2}}}{t+2} \right). \end{aligned} \tag{2.37}$$

Here we have used identity (1.45) to evaluate the sums containing H_n. The result can be simplified to

$$a_n = H_n \left((n-1)^{\underline{t}} - \frac{12}{(t+2)(t+3)} (n-3)^{\underline{t}} \right) + \frac{12(2t+5)}{(t+2)^2(t+3)^2} (n-3)^{\underline{t}}. \tag{2.38}$$

This time, when we examine the small members of the family of solutions we discover a fortunate alignment:

	x_n	a_n
$t=0$	H_n	$-H_n + \frac{5}{3}$
$t=1$	$(n-1)H_n$	$2H_n + \frac{7}{12}(n-3)$

The smallest two solutions for a_n both have leading term H_n. By an appropriate linear combination we can eliminate H_n and obtain an a_n that grows as order n:

$$x_n = (n+1)H_n \quad \leftrightarrow \quad a_n = \frac{7n+19}{12}. \tag{2.39}$$

The $s = 0$ solution from the first family is used to adjust the constant term, enabling us to reconstruct the a_n given in the original problem:

$$x_n = \tfrac{12}{7}\big((n+1)H_n + 1\big) \quad \leftrightarrow \quad a_n = n + 1. \tag{2.40}$$

This solution for x_n may not agree with the initial values x_1 and x_2. To accommodate arbitrary initial values we need to discover two extra degrees of freedom in the solution. One degree of freedom can be observed in the first family of solutions. Combining $s = 0$ with $s = 1$ yields

$$x_n = n + 1 \quad \leftrightarrow \quad a_n = 0. \tag{2.41}$$

So any multiple of $n + 1$ can be added to the solution in equation (2.40).

The second degree of freedom is not quite so obvious. Since $a_n = 0$ we have a simplified recurrence for x_n,

$$n^{\underline{3}} x_n = 12 \sum_{1<k<n} (n-k)(k-1)x_{k-1}. \tag{2.42}$$

Using a generating function, $G(z)$, for the sequence x_n, the convolution on the right of (2.42) is represented by the product of $1/(1-z)^2$ corresponding to $(n-k)$ and $G'(z)$ corresponding to $(k-1)x_{k-1}$. We obtain the differential equation

$$G'''(z) = \frac{12}{(1-z)^2} G'(z). \tag{2.43}$$

The nature of the equation suggests a solution of the form $G(z) = (1-z)^\alpha$, and testing this solution yields $\alpha = -2$ or 5. The case $\alpha = -2$ corresponds to our previous observation that multiples of $n+1$ do not affect the solution. But for $\alpha = 5$ we obtain an unusual solution that is zero after its first five values:

$$x_1 = -5, \quad x_2 = 10, \quad x_3 = -10, \quad x_4 = 5, \quad x_5 = -1. \tag{2.44}$$

This provides a second degree of freedom and gives the final solution

$$x_n = \frac{12}{7}\big((n+1)H_n + 1\big) + c_1(n+1) + c_2(-1)^n \binom{5}{n}, \tag{2.45}$$

where c_1 and c_2 are determined by the initial conditions.

2.2 Nonlinear Recurrence Relations

Nonlinear recurrence relations are understandably more difficult than their linear counterparts, and the techniques used to solve them are often less systematic, requiring conjectures and insight rather than routine tools. This section explores two types of nonlinear recurrence relations, those with maximum and minimum functions, and those with hidden or approximate linear recurrences.

2.2.1 Relations with Maximum or Minimum Functions

To solve a recurrence relation with max or min it is essential to know where the max or min occurs. This is not always obvious, since the max (or min) function may depend on earlier members of the sequence whose character is initially unknown. A typical solution strategy involves computing small values with the recurrence relation until it is possible to make a conjecture about the location of the max (or min) at each iteration. The conjecture is used to solve the recurrence and then the solution is used to prove inductively that the conjecture is correct.

This strategy is illustrated with the following example from the analysis of an *in situ* permutation algorithm [Knuth 71]. Briefly described, the problem arises in a variation of the algorithm that searches both directions simultaneously to verify cycle leaders. To check a particular j, the algorithm first looks at $p(j)$ and $p^{-1}(j)$, then at $p^2(j)$ and $p^{-2}(j)$, etc., until either encountering an element smaller than j, in which case j is not a cycle leader, or until encountering j itself, in which case j is a cycle leader since the whole cycle has been scanned.

We wish to compute the worst case cost, $f(n)$, of ruling out all the non-leaders in a cycle of size n. A recurrence arises from the observation that the second smallest element in the cycle partitions the problem. For convenience we place the cycle leader (the smallest element) at the origin and assume that the second smallest element is in the kth location.

$$\text{(leader)}\ c_1\ c_2\ c_3 \ldots c_{k-1}\ \text{(second smallest)}\ c_{k+1}\ c_{k+2} \ldots c_{n-1}. \tag{2.46}$$

Any searching among $c_1 \ldots c_{k-1}$ will not exceed the leader or the second smallest element, so the worst case for this segment is identical to the worst case for a cycle of size k. Similarly the worst for $c_{k+1} \ldots c_{n-1}$ is $f(n-k)$ and the cost of rejecting the second smallest is $\min(k, n-k)$. This gives:

$$f(n) = \max_k \big(f(k) + f(n-k) + \min(k, n-k) \big). \tag{2.47}$$

According to the strategy outlined above, our first step is to build up a table that shows the values of $f(n)$ for small n, together with a list of the values of k where the maximum is achieved:

n	$f(n)$	location of the max (k)
1	0	–
2	1	1
3	2	1, 2
4	4	2
5	5	1, 2, 3, 4
6	7	2, 3, 4
7	9	3, 4
8	12	4

In some iterations the location of the max has many possibilities, but it seems that $\lfloor n/2 \rfloor$ is always among the candidates. With the conjecture that there is a maximum at $\lfloor n/2 \rfloor$ the recurrence reduces to:

$$\begin{aligned} f(2m) &= 2f(m) + m \\ f(2m+1) &= f(m) + f(m+1) + m. \end{aligned} \tag{2.48}$$

The odd and even formulas are close enough to suggest differencing,

$$\begin{aligned} \Delta f(2n) &= f(2n+1) - f(2n) = f(n+1) - f(n) = \Delta f(n) \\ \Delta f(2n+1) &= f(2n+2) - f(2n+1) \\ &= f(n+1) - f(n) + 1 = \Delta f(n) + 1. \end{aligned} \tag{2.49}$$

In the differenced form the nature of $\Delta f(n)$ and $f(n)$ become clear: $\Delta f(n)$ simply counts the number of ones in the binary representation of n. If we let $\nu(n)$ be the number of such 1-bits then

$$f(n) = \sum_{0 \le k < n} \nu(k) = \frac{1}{2} n \log n + O(n). \tag{2.50}$$

(Digital sums like this play an important role in recurrence relations that nearly split their sequences. The asymptotic study of $f(n)$ has a confused history of independent discoveries [Stolarsky 77]. See [DeLange 75] for detailed asymptotics, and see [Knuth III; exercise 5.2.2–15] for a similar problem that depends on the binary representation of its argument.)

To complete the solution of equation (2.47) we must prove our conjecture about the location of the max, or equivalently we must show that the two-parameter function

$$g(m,n) = f(m+n) - m - f(m) - f(n), \qquad n \ge m \tag{2.51}$$

is always greater than or equal to zero. Breaking this into odd and even cases and using equation (2.48) yields

$$\begin{aligned} g(2m,2n) &= 2g(m,n) \\ g(2m+1,2n) &= g(m,n) + g(m+1,n) \\ g(2m,2n+1) &= g(m,n) + g(m,n+1) \\ g(2m+1,2n+1) &= 1 + g(m+1,n) + g(m,n+1). \end{aligned} \tag{2.52}$$

Now we can use boundary conditions that are derived from the definition of f,

$$\begin{aligned} g(n,n) &= 0 \\ g(n-1,n) &= 0, \end{aligned} \tag{2.53}$$

to prove inductively that $g(m,n) \ge 0$.

In the example above, the conjecture about the location of the maximum is straightforward and intuitive: the worst case arises when the second element is furthest from the leader so that it nearly bisects the cycle. In other examples the conjecture is more complicated. Consider the recurrence

$$f(n) = 1 + \min_k \left(\frac{k-1}{n} f(k-1) + \frac{n-k}{n} f(n-k) \right), \qquad f(1) = 0, \tag{2.54}$$

which arises from a guessing game where one player tries to determine an integer between 1 and n. After each guess the player is told whether the guess is high, low, or right on. The recurrence for $f(n)$ represents the expected number of guesses necessary by the best possible strategy.

Once again intuition tells us that it is best to choose k in the middle of the interval, but strangely enough this is not always true. The proper conjecture for locating the minimum favors dividing the interval into odd subproblems. At $n = 5$, for example, we should guess 4 rather than 3.

There are several general results that can help to locate the minimum. Included below are the first few theorems from a paper by M. Fredman and D. Knuth on recurrence relations with minimization [Fredman 74] that apply to recurrences like

$$f(n+1) = g(n+1) + \min_k \bigl(\alpha f(k) + \beta f(n-k)\bigr) \tag{2.55}$$

with α and β positive. Equation (2.54) above, when multiplied by n, is a member of this broad class.

Definition. *A real valued function $g(n)$ is convex if $\Delta^2 g(n) \geq 0$ for all n. This means that*

$$g(n+2) - g(n+1) \geq g(n+1) - g(n)\,, \qquad n \geq 0. \tag{2.56}$$

Lemma. *Let $a(n)$ and $b(n)$ be convex functions. Then the "minvolution" defined by*

$$c(n) = \min_{0 \leq k \leq n} \bigl(a(k) + b(n-k)\bigr) \tag{2.57}$$

is also convex. Moreover if $c(n) = a(k) + b(n-k)$ then

$$c(n+1) = \min\bigl(a(k) + b(n+1-k),\ a(k+1) + b(n-k)\bigr).$$

(In other words, the location of the minimum does not shift drastically as n increases. The expression "minvolution," coined by M. F. Plass, conveys the similarity of formula (2.57) to the convolution of two sequences.)

This strong lemma has a very simple proof. The process of constructing $c(n)$ can be viewed as a merging of the two sequences

$$\Delta a(0)\,,\ \Delta a(1)\,,\ \Delta a(2)\,,\ \ldots \tag{2.58}$$

and

$$\Delta b(0)\,,\ \Delta b(1)\,,\ \Delta b(2)\,,\ \ldots\,. \tag{2.59}$$

By hypothesis these two sequences are nondecreasing, so the merged sequence

$$\Delta c(0)\,,\ \Delta c(1)\,,\ \Delta c(2)\,,\ \ldots \tag{2.60}$$

is also nondecreasing, making $c(n)$ convex.

For any given n, the value of $c(n)$ is the sum of the n smallest items in the two sequences. The next value, $c(n+1)$, will require one more item from either the Δa sequence or the Δb sequence, the smaller item determining whether or not the location of the minimum shifts from k to $k+1$.

Theorem. *The function in equation* (2.55) *is convex provided that* $g(n)$ *is convex and the first iteration of the recurrence is convex:*

$$f(2) - f(1) \geq f(1) - f(0). \tag{2.61}$$

This theorem follows inductively; we assume that $f(1) \ldots f(n)$ are convex, and apply the lemma to show that $f(n+1)$ will continue the convexity.

2.2.2 Continued Fractions and Hidden Linear Recurrences

When the recurrence resembles a continued fraction, then a simple transformation will reduce the problem to a linear recurrence relation.

We consider, as an example, the problem of counting the number of trees with n nodes and height less than or equal to h, denoted by A_{nh}. For a given height h we can use the generating function

$$A_h(z) = \sum A_{nh} z^n \tag{2.62}$$

to establish a recurrence. A tree of height less than or equal to $h+1$ has a root and any number of trees of height h or less.

$$\begin{aligned} A_{h+1}(z) &= z(1 + A_h(z) + A_h(z)^2 + A_h(z)^3 + \cdots) \\ &= z/(1 - A_h(z)). \end{aligned} \tag{2.63}$$

The continued fraction flavor of this recurrence,

$$A_{h+1} = \cfrac{z}{1 - \cfrac{z}{1 - A_{h-1}(z)}}, \tag{2.64}$$

suggests the transformation

$$A_h(z) = \frac{z\, P_h(z)}{P_{h+1}(z)}, \tag{2.65}$$

which yields a linear recurrence relation:

$$P_{h+1}(z) = P_h(z) - z\, P_{h-1}(z), \qquad P_0(z) = 0, \quad P_1(z) = 1. \tag{2.66}$$

By standard techniques for quadratic linear relations we obtain

$$P_h(z) = \frac{1}{\sqrt{1-4z}}\left(\left(\frac{1+\sqrt{1-4z}}{2}\right)^h - \left(\frac{1-\sqrt{1-4z}}{2}\right)^h\right). \tag{2.67}$$

The remainder of the analysis of ordered trees, in which the coefficients of $P_h(z)$ are investigated further, does not bear directly on nonlinear recurrences, so we refer the reader to [deBruijn 72] for complete details.

It is worth noting that in seeking a transformation we were lead to a ratio of polynomials, equation (2.65), by the continued fraction nature of the recurrence. In the example above, the regularity of recurrence allowed us to use only one family of polynomials, $P_h(z)$. The underlying continued fraction theory that suggests this solution and accommodates less regular continued fractions uses two families. In general, the "nth convergent,"

$$f_n = a_0 + \cfrac{b_1}{a_1 + \cfrac{b_2}{a_2 + \ddots + \cfrac{b_n}{a_n}}}, \tag{2.68}$$

is equal to

$$f_n = p_n/q_n \tag{2.69}$$

where p_n and q_n have linear recurrence relations:

$$\begin{array}{lll} p_n = a_n p_{n-1} + b_n p_{n-2} & p_0 = a_0 & p_1 = a_1 a_0 + b_1 \\ q_n = a_n q_{n-1} + b_n q_{n-2} & q_0 = 1 & q_1 = a_1. \end{array} \tag{2.70}$$

This theory, found for example in Chapter 10 of Hardy and Wright [Hardy 79], assures us that we could reduce a less regular recurrence like

$$f_h(z) = \cfrac{z}{1 - \cfrac{z^2}{1 - f_{h-1}(z)}} \tag{2.71}$$

to a problem with two linear recurrence relations.

Besides continued fractions, there are many other types of nonlinear recurrence relations that are only thinly disguised linear recurrences. A few examples are summarized here:

Original recurrence	Linear variation
$f_n = f_{n-1} - f_n f_{n-1}$	$\frac{1}{f_{n-1}} = \frac{1}{f_n} - 1$
$f_n = f_{n-1}^3 / f_{n-2}$	$\ln f_n = 3 \ln f_{n-1} - \ln f_{n-2}$
$f_n - f_{n-1} f_n - z f_n = z - z f_{n-1}$	$f_n = \cfrac{z}{1 - \cfrac{z}{1 - f_{n-1}}}$
$f_n = 7 f_{n/2} + n^2$	$g_k = g_{k-1} + \left(\frac{4}{7}\right)^k, \quad g_k = \frac{f_{2^k}}{7^k}$

The last flavor of recurrence occurs frequently in the analysis of divide and conquer algorithms.

2.2.3 Doubly Exponential Sequences

In the preceding section we explored nonlinear recurrences that contained hidden linear relations. We turn now to a slightly different situation, where the nonlinear recurrence contains a very close approximation to a linear recurrence relation.

A surprisingly large number of nonlinear recurrences fit the pattern

$$x_{n+1} = x_n^2 + g_n, \tag{2.72}$$

where g_n is a slowly growing function of n, possibly depending on the earlier members of the sequence. As we follow the solution to (2.72) found in an article by Aho and Sloane [Aho 73], the exact requirements on g_n will become clear.

We begin by taking the logarithm of (2.72) and discovering a nearly linear formula,

$$y_{n+1} = 2y_n + \alpha_n, \tag{2.73}$$

where x_n and g_n are replaced by

$$y_n = \ln x_n; \tag{2.74}$$

$$\alpha_n = \ln\left(1 + \frac{g_n}{x_n^2}\right). \tag{2.75}$$

By using logarithms we have made our first assumption, namely that the x_n are greater than zero.

If we unroll the recurrence for y_n we obtain

$$y_n = 2^n \left(y_0 + \frac{\alpha_0}{2} + \frac{\alpha_1}{2^2} + \cdots + \frac{\alpha_{n-1}}{2^n} \right). \tag{2.76}$$

It is now convenient to extend the series in α_k to infinity:

$$Y_n = 2^n y_0 + \sum_{k=0}^{\infty} 2^{n-1-k} \alpha_k \tag{2.77}$$

$$r_n = Y_n - y_n = \sum_{k=n}^{\infty} 2^{n-1-k} \alpha_k. \tag{2.78}$$

This extension is helpful only when the series converges rapidly, so we make a second assumption: The g_n are such that

$$|\alpha_n| \geq |\alpha_{n+1}| \qquad \text{for } n \geq n_0. \tag{2.79}$$

With this second assumption Y_n is well defined and the error $|r_n|$ is bounded by the first term $|\alpha_n|$; we can exponentiate and recover the original solution:

$$x_n = e^{Y_n - r_n} = K^{2^n} \cdot e^{-r_n} \tag{2.80}$$

where

$$K = x_0 \exp\left(\sum_{k=0}^{\infty} 2^{-k-1} \alpha_k \right). \tag{2.81}$$

Since the α_k usually depend on the x_k, equation (2.80) is not a legitimate closed form solution. Nevertheless, the solution does show that there exists a constant K, perhaps hard to compute, that characterizes the sequence x_n. In some cases it is possible to determine the exact value of K.

A curious aspect of equation (2.80) is the closeness of K^{2^n} to the true solution; as we will see shortly, e^{-r_n} usually makes a negligible contribution. To demonstrate this, we will introduce a third assumption:

$$|g_n| < \tfrac{1}{4} x_n \text{ and } x_n \geq 1 \text{ for } n \geq n_0. \tag{2.82}$$

We wish to explore the closeness of $X_n = K^{2^n}$ to the exact solution x_n. Since $|r_n| \leq |\alpha_n|$, we have

$$x_n e^{-|\alpha_n|} \leq X_n \leq x_n e^{|\alpha_n|}. \tag{2.83}$$

Expanding the right side of this equation (by taking care of the case where $\alpha_n < 0$ with the identity $(1-u)^{-1} \leq 1 + 2u$ for $0 \leq u \leq 1/2$, using the third assumption) yields a new bound:

$$X_n \leq x_n + \frac{2\,|g_n|}{x_n}. \tag{2.84}$$

Similarly,

$$X_n \geq x_n e^{-|\alpha_n|} \geq x_n \left(1 - \frac{|g_n|}{x_n^2}\right) = x_n - \frac{|g_n|}{x_n}. \tag{2.85}$$

Finally, the assumption $|g_n| < \frac{1}{4} x_n$ permits us to claim that

$$|x_n - X_n| < \frac{1}{2}. \tag{2.86}$$

So in cases where we know that x_n is an integer the solution is

$$x_n = \text{ nearest integer to } K^{2^n}, \quad \text{for } n \geq n_0. \tag{2.87}$$

Here are several recurrence relations that fit the general pattern given by equation (2.72):

1) Golomb's Nonlinear Recurrences.

$$y_{n+1} = y_0\, y_1\, \ldots\, y_n + r, \qquad y_0 = 1. \tag{2.88}$$

This definition is equivalent to the finite-history recurrence

$$y_{n+1} = (y_n - r)y_n + r, \qquad y_0 = 1, \quad y_1 = r + 1. \tag{2.89}$$

And when the square is completed with the following substitution

$$x_n = y_n - \frac{r}{2} \tag{2.90}$$

$$x_{n+1} = x_n^2 + \frac{r}{2} - \frac{r^2}{4} \tag{2.91}$$

the recurrence becomes an obvious member of the family just solved. Since the g_n term is constant, it is easy to verify that all the assumptions are satisfied.

In the special cases $r = 2$ and $r = 4$, the constant k is known to be equal to $\sqrt{2}$ and the golden ratio respectively. In other cases the constant can be estimated by iterating the recurrence and solving for k. The doubly exponential growth of the sequence makes such estimates converge rapidly; but it also makes the estimates inexact for further terms in the sequence.

2) Balanced Trees.

The following recurrence, given in [Knuth III; Section 6.2.3], counts the number of balanced binary trees of height n.

$$y_{n+1} = y_n^2 + 2y_n y_{n-1}. \tag{2.92}$$

When we make the transformation $x_n = y_n + y_{n-1}$ the recurrence appears in a more complex yet more tractable form,

$$x_{n+1} = x_n^2 + 2y_{n-1}y_{n-2}. \tag{2.93}$$

Here the g_n term is not constant, but grows slowly enough ($2y_{n-1}y_{n-2} \ll y_n < x_n$) to meet the requirements on g_n. We can assert that there exists a k such that

$$x_n = \lfloor k^{2^n} \rfloor \tag{2.94}$$

and

$$y_n = \lfloor k^{2^n} \rfloor - \lfloor k^{2^n - 1} \rfloor + \cdots \pm \lfloor k \rfloor \pm 1. \tag{2.95}$$

(The use of the floor function in place of the nearest integer is a consequence of g_n being positive, making k^{2^n} always slightly larger than the correct value.)

We conclude with two recurrences mentioned in Aho and Sloane:

$$y_{n+1} = y_n^3 - 3y_n \tag{2.96}$$

$$y_{n+1} = y_n y_{n-1} + 1. \tag{2.97}$$

Strictly speaking, these relations do not fit the pattern solved at the beginning of this section. However, the techniques developed earlier are equally applicable. After taking logarithms both recurrences become nearly linear. Equation (2.97), for example, has a Fibonacci-like solution:

$$y_n = \lfloor k_0^{F_{n-1}} k_1^{F_n} \rfloor, \tag{2.98}$$

where $F_n = F_{n-1} + F_{n-2}$.

Chapter 3

Operator Methods

The following analysis of hashing, based on unpublished notes by Michael Paterson, relies on two concepts: eigenoperators and what he calls "induction from the other end." The cookie monster example below illustrates the value of finding an eigenoperator. "Induction at the other end" will appear later when we apply the techniques to various hashing schemes.

3.1 The Cookie Monster

Consider a monster whose ability to catch cookies is proportional to its current size. When we throw a cookie and the monster has k cookies in its belly, with probability pk the monster grows to size $k+1$. (We assume that $pk \leq 1$.)

Let g_{nk} be the probability that the monster is size k after n cookies have been thrown. We construct a generating function,

$$g_n(x) = \sum_k g_{nk}\, x^k, \tag{3.1}$$

that represents the distribution after n cookies. Initially the monster has somehow eaten one cookie, that is, $g_0(x) = x$.

Given $g_n(x)$, an "operator" will provide a means of obtaining $g_{n+1}(x)$. First let's look at the impact of a cookie on an individual term.

Before:	After:
x^k	$pk\, x^{k+1} + (1 - pk)\, x^k$
	or $x^k + p(x-1)k\, x^k$

This change is captured by the operator $\Phi = 1 + p\,(x-1)xD$, where D is the derivative. Applying Φ repeatedly gives $g_n(x) = \Phi^n g_0(x)$.

Before proceeding further it is helpful to review some facts about operators. We will be using the following:

D	derivative
U	evaluate at $x = 1$
Z	evaluate at $x = 0$
UD	obtain the mean $f'(1)$
U_n	shorthand for UD^n
x^n	multiply by x^n

It will be important to understand how operators commute with one another. For example,

$$D\,x^n\,f(x) = n\,x^{n-1} f(x) + x^n\,D\,f(x), \tag{3.2}$$

so we can move D past x^n by the formula:

$$Dx^n = x^n D + n\,x^{n-1}. \tag{3.3}$$

This generalizes to arbitrary polynomials $r(x)$:

$$D\,r(x) = r(x)\,D + r'(x). \tag{3.4}$$

Another useful fact about operators is the relation

$$U_n x = U_n + n\,U_{n-1} \tag{3.5}$$

or

$$U_n(x-1) = n\,U_{n-1}. \tag{3.6}$$

This can be shown by commuting x with each of the D operators in $U_n x$.

Returning to the cookie monster, we would like to obtain the mean size of the monster after n cookies:

$$U_1\,g_n(x) = U_1\,\Phi^n\,g_0(x). \tag{3.7}$$

Here is where commuting is important, since it would be nice to be able to move U_1 past Φ. Applying U_1 to Φ gives

$$\begin{aligned} UD\,\Phi &= UD(1 + p(x-1)xD) \\ &= U\left(D + p(x-1)xD^2 + p(2x-1)D\right) \\ &= (1+p)\,UD. \end{aligned} \tag{3.8}$$

So UD is an eigenoperator of Φ, and by this self-replication we can compute the mean:

$$U_1\Phi^n g_0(x) = (1+p)^n U_1 g_0(x) = (1+p)^n. \tag{3.9}$$

The variance is obtained with U_2, since $\mathrm{Var}(g) = g''(1) + g'(1) - (g'(1))^2$ and $U_2 g_n(x) = g_n''(1)$. Unfortunately, U_2 does not have the nice eigenoperator property that U_1 possesses; we have

$$\begin{aligned} U_2\Phi &= U(D^2 + p(x-1)xD^3 + 2p(2x-1)D^2 + 2pD) \\ &= UD^2 + 2pUD^2 + 2pUD \\ &= (1+2p)U_2 + 2pU_1. \end{aligned} \tag{3.10}$$

However, by a suitable linear combination with U_1, we do obtain an eigenoperator:

$$(U_2 + 2U_1)\Phi = (1+2p)(U_2 + 2U_1). \tag{3.11}$$

In fact there is a whole family of eigenoperators given by the scheme

$$\begin{array}{ll} V_1\Phi = (1+p)V_1 & V_1 = U_1 \\ V_2\Phi = (1+2p)V_2 & V_2 = U_2 + 2U_1 \\ V_3\Phi = (1+3p)V_3 & V_3 = U_3 + 6U_2 + 6U_1 \\ V_n\Phi = (1+np)V_n & V_n = U_n x^{n-1}. \end{array}$$

This can be shown with equations (3.6) and (3.3):

$$\begin{aligned} V_n\Phi &= U_n x^{n-1}(1 + p(x-1)x\,D) \\ &= V_n + U_n p(x-1)x^n D \\ &= V_n + pnU_{n-1}(Dx^n - nx^{n-1}) \\ &= V_n + pn(U_n x - n\,U_{n-1})x^{n-1} \\ &= V_n + pnV_n. \end{aligned} \tag{3.12}$$

In principle we can therefore recover all the higher moments of the distribution using the V_i. The variance, for example, is computed with V_2:

$$V_2\Phi^n g_0(x) = (1+2p)^n V_2 x = 2(1+2p)^n$$

$$\begin{aligned} U_2\Phi^n g_0(x) &= (V_2 - 2V_1)\Phi^n g_0(x) \\ &= 2(1+2p)^n - 2(1+p)^n \end{aligned}$$

$$\begin{aligned} \mathrm{Var}(g_n) &= g_n''(1) + g_n'(1) - (g_n'(1))^2 \\ &= 2(1+2p)^n - (1+p)^n - (1+p)^{2n}. \end{aligned} \tag{3.13}$$

3.2 Coalesced Hashing

A moment's reflection indicates that the behavior of the cookie monster is very closely related to certain kinds of hashing. When keys collide, a long chain develops, and the likelihood of hitting the chain increases. Suppose we resolve collisions by finding the first free spot at the left end of the table and by linking this spot on the end of the chain. As the algorithm proceeds we will have a distribution of monsters of various sizes. Let

$$g_n(x) = \sum_k (\text{expected number of chains of length } k)\, x^k. \tag{3.14}$$

Once again we would like to find an operator that describes the addition of a cookie, but this time we think of keys instead of cookies.

Because we are dealing with expected values, the general term will behave like a single cookie monster, even though there may be several monsters involved. Here $p = 1/m$, where m is the number of slots in the hash table. So the probability of a chain of length k growing to $k+1$ is pk, the probability of hitting the chain. However, the computation of the constant term in the generating function presents new difficulties. The expected number of empty chains is just $m - n$, so the operator must be:

$$\Psi = \Phi + (\text{fudge the constant term to } m - n). \tag{3.15}$$

Without fudging, Φ applied to the constant term of $g_n(x)$ is $\Phi(m-n) = m - n$. The correct change should be:

Before:	After:
$m - n$	$(m - n - 1) + (1 - np)\,x$

We can patch Φ using the evaluate-at-zero operator, Z:

$$\Psi = \Phi + p(x-1)Z - pU_1. \tag{3.16}$$

Note that Z applied to $g_n(x)$ gives $m - n$ and U_1 gives n, so Ψ performs properly on the constant term:

$$\begin{aligned} \Psi(m-n) &= m - n + p(x-1)(m-n) - pn \\ &= m - n - mp + (1 - np)\,x. \end{aligned} \tag{3.17}$$

(Recall that $mp = 1$.) Using Z and U_1 for this fudge might seem at first like a difficult way of accomplishing a simple fix, but it is important that the change be done entirely with linear operators.

Now $g_n(x)$ is given by

$$g_n(x) = \Psi^n g_0(x), \qquad g_0(x) = m. \tag{3.18}$$

As before, we seek an eigenoperator of Ψ; the application of U_1 to Ψ gives

$$U_1\Psi = (1+p)U_1 + pZ. \tag{3.19}$$

There is no systematic way to find eigenoperators, but the presence of Z suggests trying

$$Z\,\Psi = (1-p)\,Z - p\,U_1. \tag{3.20}$$

We see now that the following linear combination is an eigenoperator:

$$(U_1 + Z)\Psi = (U_1 + Z). \tag{3.21}$$

The "mean" in this problem is not particularly interesting, since U_1 applied to $g_n(x)$ is just n, and the eigenoperator confirms this fact:

$$(U_1 + Z)g_n(x) = 1^n m; \tag{3.22}$$

$$Zg_n(x) = m - n. \tag{3.23}$$

The power of the eigenoperator lies instead in the computation of the expected number of collisions on the $(n+1)$st insertion. Let

$$h_n(x) = \sum_k \bigl(\text{probability of } k \text{ collisions on the } (n+1)\text{st insertion}\bigr)\, x^k. \tag{3.24}$$

The x^k term in $g_n(x)$ will contribute $(x^k + x^{k-1} + \cdots + x)p$ to $h_n(x)$, because each item in a k-chain is equally likely to be hit, yet they are at different distances from the end of the chain.

We want to compute $U_1 h_n(x)$ based on $g_n(x)$. Applying U_{r+1} to a polynomial, and taking liberties with the constant term, gives

$$\begin{aligned} U_{r+1}(x^{k+1}) &= U_{r+1}(x^{k+1} - 1) \\ &= U_{r+1}(x-1)(1 + x + \cdots + x^k) \\ &= (r+1)U_r(1 + x + \cdots + x^k) \\ &= (r+1)U_r(x + x^2 + \cdots + x^k). \end{aligned} \tag{3.25}$$

(These liberties are justified because we are applying U to the polynomial argument x^{k+1}; we are not commuting U with the operator x^{k+1} as in equation (3.3).) Using $r = 1$ relates g and h,

$$U_1 h_n(x) = \frac{p}{2} U_2 x g_n(x). \tag{3.26}$$

Since $U_2 x = U_2 + 2U_1$, and since U_1 is easy to compute, we now seek an eigenoperator of Ψ that contains U_2. Here is an appropriate family of eigenoperators:

$$\begin{aligned} C_2\Psi &= (1+2p)C_2 & C_2 &= V_2 - \tfrac{1}{2}(U_1 - Z) \\ C_3\Psi &= (1+3p)C_3 & C_3 &= V_3 - \tfrac{2}{3}(U_1 - 2Z) \\ C_n\Psi &= (1+np)C_n & C_n &= V_n - \tfrac{(n-1)!}{n}(U_1 - (n-1)Z). \end{aligned}$$

This enables us to find all the higher moments of the distribution of collisions necessary to insert the $(n+1)$st element. For instance, the mean number of collisions is obtained with the C_2 operator:

$$\begin{aligned} U_1 h_n(x) &= \frac{p}{2}(U_2 + 2U_1)\, g_n(x) \\ &= \frac{p}{2}\left(C_2 + \frac{U_1}{2} - \frac{Z}{2}\right) g_n(x) \\ &= \frac{1}{2m}\left(\left(1 + \frac{2}{m}\right)^n \frac{m}{2} - \frac{m}{2} + n\right). \end{aligned} \tag{3.27}$$

The reader might have noticed that the last analysis takes no account of the time necessary to find the first free cell on the left end of the array. Suppose that the hashing algorithm uses a pointer to keep track of the previous free cell. After each collision the pointer is moved rightward until a new free cell is discovered. The algorithm is modeled by the following game. We start with an empty array and a pointer at zero. The game requires n "R-steps," after which we compute the distance from the pointer to the next free cell. When there are j unoccupied cells, an "R-step" occupies an empty cell with probability pj or occupies the leftmost free cell with probability $(1 - pj)$. The second case corresponds to a collision, and the pointer is set to the recently occupied cell. The final score of the game, the distance between the pointer and the next free cell, gives the cost of finding an empty cell for a future collision.

Once again we use a generating function. Let $G_{mn}(z)$ be

$$\sum_k (\text{probability that the score is } k \text{ in an } m \text{ array after } n \text{ R-steps})\, z^k. \tag{3.28}$$

We seek an operator to construct G_{mn} from smaller problems, this time with a different style of induction. Suppose we have a sequence of R-steps:

$$3 \quad 1 \quad 4 \quad C \quad 7$$

The numbers indicate cells occupied, and C represents a collision where the leftmost free cell is occupied and the pointer adjusted. Every such sequence of steps has a certain probability of occurring, and leads to a certain score, as defined above.

Rather than add a new element to the end of the sequence we place it at the beginning, hence the expression "induction at the other end." Specifically, we will add a new key and a new cell to the array. The key can fit anywhere in the old array, so we can describe it as the addition of $k \in \{\frac{1}{2}, \frac{3}{2}, \ldots, \frac{2m+1}{2}, C\}$ at the beginning of the sequence, and a renumbering to make the sequence integer again.

For example, consider the R-steps given above, and assume that the array size is $m = 7$. When the C arrives cell 1 is occupied, so it lands in cell 2. At the end of the game the next free cell is 5, so the score is 3. Here are the possible changes, depending which new R-step we place at the beginning of the sequence:

Probability:	New First Element:	Remaining Sequence:	Score:
p	1	$4\,2\,5\,C\,8$	3
p	2	$4\,1\,5\,C\,8$	3
p	3	$4\,1\,5\,C\,8$	4
p	4	$3\,1\,5\,C\,8$	4
p	5	$3\,1\,4\,C\,8$	4
p	6	$3\,1\,4\,C\,8$	3
p	7	$3\,1\,4\,C\,8$	3
p	8	$3\,1\,4\,C\,7$	3
$1-8p$	C	$4\,2\,5\,C\,8$	3

How will this affect the final score? The score is the length of the region between the pointer and the next free cell. If the new key lands in this region the score is increased by one, otherwise the score remains unchanged. Since the probability of hitting this region is proportional to the region size, the cookie monster rears his ugly head and with the familiar Φ operator he devours the rest of this analysis:

$$G_{mn}(x) = \Phi\, G_{m-1,n-1}(x) = \Phi^n x. \tag{3.29}$$

([Knuth III; exercise 6.4–41] has a less elegant solution to this problem, and says, "Such a simple formula deserves a simpler proof!")

3.3 Open Addressing: Uniform Hashing

Let us consider a slight variation on the previous game. Instead of an R-step we use a T-step that fills an empty cell at random and leaves the pointer at the left end of the array. The final score is the distance from the left end of the array to the first free cell.

Motivation for this new game comes from the slightly unrealistic assumption that each key has a random permutation for a probe sequence. The key pursues its probe sequence until it finds an empty cell. This assumption, usually called uniform hashing, will be refined later when we discuss secondary clustering.

We would like to determine the expected number of entries that the $(n + 1)$st element must examine in its probe sequence. We are free to assume that this element has $1, 2, 3, \ldots$ for a probe sequence by rearranging the array if necessary so that this is true. Then the $(n + 1)$st insertion requires finding the leftmost free cell, and this is equal to the score of the T-step game described above.

Using induction at the other end, we run into the cookie monster once again. This time he has occupied the cells at the beginning of the array. However, we must be careful about the probability p. The probability of landing in a given cell is $1/m$, so the operator is

$$\Phi_m = 1 + \frac{1}{m}(x - 1)xD. \tag{3.30}$$

Remember that induction at the other end adds both a new key and a new array slot, so that the probability changes and we must parameterize the operator Φ with m.

With this parameterized operator, the generating function for monster size is given by

$$G_{mn}(x) = \Phi_m \Phi_{m-1} \ldots \Phi_{m-n+1} x. \tag{3.31}$$

V_1 and V_2 are still eigenoperators; they give products that telescope nicely. For example, the average number of probes used to insert the $(n + 1)$st element is

$$\begin{aligned} V_1 G_{mn}(x) &= \left(1 + \frac{1}{m}\right)\left(1 + \frac{1}{m-1}\right) \ldots \left(1 + \frac{1}{m-n+1}\right) \\ &= \frac{m+1}{m-n+1}. \end{aligned} \tag{3.32}$$

And since all the V_i telescope there is a systematic way of computing the mean and variance of the probes necessary to insert the $(n+1)$st element.

3.4 Open Addressing: Secondary Clustering

In the secondary clustering model each key is mapped to a single hash value, then the hash value provides a random permutation for the probe sequence. Rather than each key having its own random probe sequence, the keys share probe sequences with those keys mapping to the same hash value. The hash values and the probe sequences are still random, but the additional sharing makes collisions more likely.

This time the game we play has an S-rule: If the leftmost cell is unoccupied use rule S0 otherwise use S1. Rule S0 occupies an empty cell at random. Rule S1 has a choice: With probability p it occupies the leftmost empty cell, and with probability $q = 1 - p$ it occupies any empty cell at random.

The S-rule captures the somewhat subtle behavior of secondary clustering. We assume without loss of generality that each key hashing to the leftmost cell has probe sequence 1, 2, 3, ... Then in rule S1 with probability p we hash to the leftmost cell, reuse the same hash sequence 1, 2, 3, ..., and occupy the leftmost empty cell.

The score is the distance to the first free cell, and we have two score-generating functions for the two rules: $H_{mn}(x)$ for S0 and $G_{mn}(x)$ for S1. Let's look first at $G_{mn}(x)$:

$$G_{mn}(x) = (px + q\Phi_m)\, G_{m-1,n-1}(x). \tag{3.33}$$

The operator for G is derived as before by using "induction from the other end." With probability p the key lands at location one and increases the monster by one. With probability q we play the old cookie monster game by adding a key at random.

There is a fine distinction among probabilities in this operator: The probability p is fixed at $1/m$ before the induction step and remains fixed as m decreases. The operator Φ_m, however, is parameterized with m, so the probability in this operator increases with smaller m. The distinction is precisely what we want, since the probability of a new key sharing the same probe sequence with a particular old key is fixed at $1/m$ throughout the process.

The quantity $\Omega_m = px + q\Phi_m$ in (3.33) does not have an eigenoperator, but it does have a "sliding" operator:

$$(U_1 - (m+1)\,U_0)\,\Omega_m = (1 + \frac{q}{m})(U_1 - m\,U_0). \tag{3.34}$$

The sliding operator $A_m = U_1 - m\,U_0$ changes its parameter by one when it commutes with Ω_m, and this behavior is just as valuable as an eigenoperator when we want to compute U_1:

$$A_{m+1}G_{mn}(x) = \left(1 + \frac{q}{m}\right)\left(1 + \frac{q}{m-1}\right) \cdots \left(1 + \frac{q}{m-n+1}\right) A_{m-n+1}x \tag{3.35}$$

$$U_1 G_{mn}(x) = \left(1 + \frac{q}{m}\right)\left(1 + \frac{q}{m-1}\right) \cdots \left(1 + \frac{q}{m-n+1}\right)(n-m) + (m+1) \tag{3.36}$$

Now we can turn our attention to $H_{mn}(x)$ and rule S0. Until the first cell is occupied this also behaves like a cookie monster. Once the first cell is hit, we switch to $G_{mn}(x)$. Using induction at the other end, this gives the recurrence:

$$H_{mn}(x) = \Phi_m H_{m-1,n-1}(x) - \frac{x}{m} H_{m-1,n-1}(x) + \frac{x}{m} G_{m-1,n-1}(x). \tag{3.37}$$

The middle term corresponds to a mistaken use of H by the Φ_m operator in the case of an occupied first cell.

Since the game begins in S0, H_{mn} is the desired generating function for the whole game, and we would like to find its mean, $U_1 H_{mn}(x)$:

$$\begin{aligned} U_1 H_{mn}(x) &= \left(1 + \frac{1}{m}\right) U_1 H_{m-1,n-1} - \frac{U_1 x}{m} H_{m-1,n-1} + \frac{U_1 x}{m} G_{m-1,n-1} \\ &= U_1 H_{m-1,n-1} + \frac{U_1}{m} G_{m-1,n-1}. \end{aligned} \tag{3.38}$$

A similar recurrence for G_{mn} can be deduced from equation (3.34):

$$U_1 G_{mn}(x) = \left(1 + \frac{q}{m}\right) U_1 G_{m-1,n-1} + p. \tag{3.39}$$

The situation calls for a new operator trick. Note that a linear combination of H and G replicates itself:

$$U_1\left(H_{mn} - \frac{1}{q}G_{mn}\right) = U_1\left(H_{m-1,n-1} - \frac{1}{q}G_{m-1,n-1}\right) - \frac{p}{q}. \qquad (3.40)$$

Furthermore, the term $\frac{p}{q}$ is independent of m, so we have

$$U_1\left(H_{mn} - \frac{1}{q}G_{mn}\right) = U_1\left(H_{m-n,0} - \frac{1}{q}G_{m-n,0}\right) - \frac{np}{q}. \qquad (3.41)$$

Given the boundary conditions $H_{m,0} = G_{m,0} = x$ and the previously computed $U_1 G_{mn}$, we can determine $U_1 H_{mn}$:

$$\begin{aligned} U_1 H_{mn} &= U_1\left(H_{m-n,0} - \frac{1}{q}G_{m-n,0} + \frac{1}{q}G_{mn}\right) - \frac{np}{q} \\ &= 1 + \frac{1}{q}\left(m - np + (n-m)\prod_{k=m-n+1}^{m}\left(1+\frac{q}{k}\right)\right). \end{aligned} \qquad (3.42)$$

(It is interesting to compare the solution above with the brute-force approach to hashing found in [Knuth III; exercise 6.4–44].)

The last operator trick bears a strong resemblance to the earlier use of eigenoperators and sliding operators. In all of these cases we moved through the recurrence by a self-replicating process. The power of operator methods lies in their ability to hide unimportant details so that this kind of self-replication becomes apparent; therefore quantities like means and variances become relatively easy to compute.

Chapter 4

Asymptotic Analysis

4.1 Basic Concepts

There is no guarantee that the study of algorithms will produce sums and recurrences with straightforward closed form solutions. In fact much of the adventure of analysis of algorithms lies in the variety of mathematics to which researchers are drawn (at times kicking and screaming) in their attempts to understand algorithms. Frequently the researchers will turn to asymptotic analysis.

Asymptotic analysis attempts to find a solution that closely approximates the exact solution. Often the relative error of this approximation becomes small for large values of the parameters involved. We will attempt to discover as thorough an asymptotic approximation as possible. For example, instead of knowing that an algorithm runs in $O(n^2)$ time it will be far more satisfying to know that the running time is $3n^2 + 7n + O(1)$.

Giving attention to asymptotic detail has several rewards. Frequently the approximate solution converges so rapidly that the researcher can test a few small cases and have immediate confirmation of the correctness of a solution. It is important in practice to know more than the leading term, since $1.8 \ln n + 20$ will be smaller than $2 \ln n + 10$ only when $n > e^{50}$. Moreover, the pursuit of additional asymptotic terms usually leads to more general and powerful mathematical techniques.

The purpose of this chapter is to introduce the basic tools of asymptotics: O-notation, bootstrapping, and dissecting. The first few sections will describe these ideas briefly, and the last section includes the derivation of an asymptotic result that is difficult as a whole, but basic at each step.

4.1.1 Notation

Definition of O or $\preceq$.

We say that $f(n) = O(g(n))$ (or $f(n) \preceq g(n)$) as $n \to \infty$ if there exist integers N and K such that $|f(n)| \leq K\,|g(n)|$ for all $n \geq N$.

Definition of Ω or $\succeq$.

In a similar vein, $f(n) = \Omega(g(n))$ (or $f(n) \succeq g(n)$) as $n \to \infty$ if there exist integers N and K such that $|f(n)| \geq K\,|g(n)|$ for all $n \geq N$.

When both of these definitions apply, the situation is denoted by $f(n) = \Theta(g(n))$ or $f(n) \asymp g(n)$ [Knuth 76b].

There are similar definitions for little o notation. For example, $f(n) = o(g(n))$ or $f(n) \prec g(n)$ whenever $\lim_{n\to\infty} f(n)/g(n) = 0$. There is also a notation for equivalence, $f(n) \sim g(n)$ if $\lim_{n\to\infty} f(n)/g(n) = 1$. However, in general we will avoid these notations because they do not capture information about the rate of convergence of the limits involved. We prefer to use a strong assertion like $O(n^{-1/2})$ instead of a weak one like $o(1)$.

4.1.2 Bootstrapping

Bootstrapping is helpful in situations where there is an implicit equation for a given function of interest. By repeatedly feeding asymptotic information about the function back into the equation the approximation is steadily improved. Here is an example from [deBruijn 70]:

$$f(t)e^{f(t)} = t, \qquad t \to \infty. \tag{4.1}$$

The formula can be rewritten as

$$f(t) = \ln t - \ln f(t). \tag{4.2}$$

We "prime the pump" by observing that for $t > e$ we have $f(t) > 1$. Using this in equation (4.2) gives

$$f(t) = O(\ln t). \tag{4.3}$$

Inserting the approximation again into (4.2) yields a better result:

$$f(t) = \ln t + O(\ln\ln t). \tag{4.4}$$

Once again we feed this result back into equation (4.2) to improve the result further,

$$f(t) = \ln t - \ln\ln t - \ln\left(1 + O\left(\frac{\ln\ln t}{\ln t}\right)\right)$$
$$= \ln t - \ln\ln t + O\left(\frac{\ln\ln t}{\ln t}\right). \tag{4.5}$$

In this manner the approximation can be bootstrapped to any degree of accuracy.

4.1.3 Dissecting

Dissecting is applied chiefly to sums and integrals. In a typical situation a sum is given over a large range, and the summand has several components. No single component of the summand is small throughout the range, but if the range is dissected into pieces then each piece becomes small (for a variety of different reasons) and in this fashion the whole sum is shown to be small.

The dissection technique can be illustrated by the sum

$$f(n) = \sum_{3 \le d \le n/2} \frac{1}{d\,(n/d)^d}. \tag{4.6}$$

We break the sum into three intervals. When $3 \le d \le 8$ the sum is less than

$$\sum_{3 \le d \le 8} \frac{1}{3(n/8)^3} = O(n^{-3}). \tag{4.7}$$

Note that the d's in the original formula are replaced by 3 or 8 in equation (4.7), depending on their worst possible effect on the sum. Then the constant number of terms in the sum allows us to claim a $O(n^{-3})$ bound.

On the second interval, $8 \le d \le \sqrt{n}$, we do a similar replacement of d by its extreme values so that the sum is less than

$$\sum_{8 \le d \le \sqrt{n}} \frac{1}{8(n/\sqrt{n}\,)^8} = O(n^{-4}\sqrt{n}\,). \tag{4.8}$$

Here the $O(n^{-4}\sqrt{n}\,)$ bound is caused by $O(\sqrt{n}\,)$ terms of size at most $O(n^{-4})$.

The sum over the remaining interval, $\sqrt{n} \le d \le n/2$, is extremely small, since it is less than

$$\sum_{\sqrt{n}\le d\le n/2} \frac{1}{\sqrt{n}\,2^{\sqrt{n}}} = O\left(\frac{\sqrt{n}}{2^{\sqrt{n}}}\right). \tag{4.9}$$

Combining the three intervals, we conclude that the whole sum is $O(n^{-3})$.

It is clear from the example above that the difficulty of dissecting lies in the choice of intervals. The division points 8 and $\sqrt{n}$ are not sacred: 10 and $\sqrt[3]{n}$, for example, work equally well. Nevertheless the choice of 8 and $\sqrt{n}$ is somewhat of an art requiring insight into the behavior of the summand throughout the entire interval.

4.1.4 Limits of Limits

Occasionally an asymptotic argument will involve two or more limiting processes. The ordering of the limits is often critical, and it is useful to know when the exchange of limits is permissible. In simple situations like

$$\sum_{n=0}^{\infty}\sum_{m=0}^{\infty} a_{mn} \tag{4.10}$$

the absolute convergence of the a_{mn} allows the series to be rearranged at will. We could, for example, sum on n before m.

Later in this chapter we need to change limits in more delicate circumstances. In particular, we want to invert the following theorem:

Abelian Theorem. *If*

$$\lim_{n\to\infty} \sum_{k=0}^{n} a_k = A$$

then

$$\lim_{z\to 1-}\ \lim_{n\to\infty} \sum_{k=0}^{n} a_k z^k = A.$$

(In this limit and hereafter we assume that z approaches unity from below.)

The converse statement is not always true:

False Conjecture. *If*

$$\lim_{z\to 1-}\lim_{n\to\infty}\sum_{k=0}^{n} a_k z^k = A$$

then

$$\lim_{n\to\infty}\sum_{k=0}^{n} a_k = A.$$

N. G. de Bruijn gives the following counterexample. Let

$$f(z) = \frac{1-z}{1+z} = 1 - 2z + 2z^2 - 2z^3 + \cdots \tag{4.11}$$

and let a_k be the coefficients of the power series expansion for $f(z)$. The series converges absolutely within a circle of radius one around the origin, and its limit at one is zero:

$$\lim_{z\to 1-} f(z) = 0. \tag{4.12}$$

But the partial sums of a_k will never converge to zero:

$$a_0 + a_1 + \cdots + a_n = (-1)^n. \tag{4.13}$$

Tauber supplied an additional requirement to invert Abel's theorem. He stipulated that a_k must be $o(k^{-1})$. Hardy and Littlewood subsequently weakened this condition to $a_k > -C\,k^{-1}$ for some $C > 0$, although the theorem is still labeled Tauberian because of the general flavor of the result. Tauberian theorems supply the conditions necessary to invert Abelian theorems.

Tauberian Theorem. *If*

$$\lim_{z\to 1-}\lim_{n\to\infty}\sum_{k=0}^{n} a_k z^k = A$$

and if $a_k > -Ck^{-1}$ *for some* $C > 0$, *then*

$$\lim_{n\to\infty}\sum_{k=0}^{n} a_k = A.$$

For a collection of deeper Tauberian theorems see [Hardy 49; page 154].

4.1.5 Summary of Useful Asymptotic Expansions

In the formulas below, n tends to infinity and ϵ tends to zero.

$$H_n = \ln n + \gamma + \frac{1}{2n} - \frac{1}{12n^2} + O(n^{-4}) \qquad (4.14)$$

$$n! = \sqrt{2\pi n}\left(\frac{n}{e}\right)^n \left(1 + \frac{1}{12n} + \frac{1}{288n^2} + O(n^{-3})\right) \qquad (4.15)$$

$$\ln(1+\epsilon) = \epsilon - \frac{\epsilon^2}{2} + \frac{\epsilon^3}{3} - \frac{\epsilon^4}{4} + \cdots + (-1)^{m-1}\frac{\epsilon^m}{m} + O(\epsilon^{m+1}) \qquad (4.16)$$

$$\begin{aligned}\sum_{k=1}^{n} k^m &= \frac{B_{m+1}(n) - B_{m+1}}{m+1} \qquad \text{integer } m, n > 0 \\ &= \frac{n^{m+1}}{m+1} + \frac{n^m}{2} + \frac{mn^{m-1}}{12} + O(n^{m-2}), \qquad m > 1\end{aligned} \qquad (4.17)$$

($B_i(x)$ and B_i are the Bernoulli polynomials and numbers, see page 59.)

$$\sum_{k=n_0}^{n} \frac{1}{k \ln k \ln\ln k \ldots \left(\ln^{(i)} k\right)^{1+\epsilon}} = O(1), \qquad \epsilon > 0 \qquad (4.18)$$

The last equation represents the turning point for sums. When $\epsilon = 0$ the sums will diverge. For example, the sums

$$\sum \frac{1}{k}, \quad \sum \frac{1}{k \ln k}, \quad \text{and} \quad \sum \frac{1}{k \ln k \ln\ln k}$$

are all unbounded.

There are several ways to obtain crude estimates. One involves the replacement of sums by their integral counterparts. In Section 4.2.2 on Euler's summation formula we will see when this substitution is valid, and how to refine the results of the approximation. Another estimate applies to random variables with mean μ and variance σ^2. Chebyshev's inequality tells us that

$$\mathrm{Prob}\left(|X - \mu| \geq t\right) \leq \frac{\sigma^2}{t^2}. \qquad (4.19)$$

In Section 4.3.3 we will develop detailed formulas for the case where X is a sum of independent random variables.

4.1.6 An Example from Factorization Theory

We turn now to the problem of computing the probability that a polynomial of degree n has irreducible factors of distinct degrees modulo a large prime p, a situation that is advantageous for certain factoring algorithms [Knuth II; pages 429–431]. The probability that an nth degree polynomial is itself irreducible mod p is

$$\frac{1}{n} + O(p^{-n/2}). \tag{4.20}$$

(This result is proved, for example, in [Knuth II; exercise 4.6.2-4].) The modulus, p, is unimportant, so we let p go to infinity and use probability $1/n$ as a foundation for the more difficult problem of factoring into distinct-degree polynomials.

The solution relies on a partition-style generating function. The coefficient of z^n in

$$h(z) = \prod_{k\geq 1} \left(1 + \frac{z^k}{k}\right) \tag{4.21}$$

is the desired solution, that is, the probability of a distinct-degree factorization. To see this, note that if h_n is the coefficient of z^n, h_n will be a sum of terms like

$$\frac{z^{k_1}}{k_1}\frac{z^{k_2}}{k_2}\cdots\frac{z^{k_m}}{k_m} \tag{4.22}$$

where each of the k's is distinct. Each term like (4.22) corresponds to a partition of n into distinct integers $k_1, k_2, \ldots, k_m$. Suppose we are to construct a polynomial of size n by multiplying polynomials of sizes $k_1, k_2, \ldots, k_m$. (We assume that these small polynomials and the large polynomial are all monic. Other leading coefficients do not affect the results that follow.) There are p^{k_1} polynomials of degree k_1. Of these p^{k_1}/k_1 are irreducible, by our assumption. Treating each polynomial this way gives a total of

$$\frac{p^{k_1}}{k_1}\frac{p^{k_2}}{k_2}\cdots\frac{p^{k_m}}{k_m} = \frac{p^n}{k_1\, k_2 \ldots k_m} \tag{4.23}$$

polynomials whose irreducible factors have the appropriate sizes. Since there are a total of p^n monic polynomial of size n, this means that the coefficient

$$\frac{1}{k_1\, k_2 \ldots k_m} \tag{4.24}$$

in equation (4.22) is the probability of obtaining a factorization into irreducible parts of distinct sizes $k_1, k_2, \ldots, k_m$.

The whole of h_n consists of all possible partitions, each contributing a term of the form (4.22), and since all of the events are disjoint these probabilities are summed. Thus the generating function properly determines h_n, the limiting probability that a polynomial of degree n factors into irreducible parts of distinct sizes modulo a large prime.

Equation (4.21) does not give us a closed form for h_n, and there does not seem to be one, so instead we seek an asymptotic formula as $n \to \infty$. Taking logarithms and expanding each logarithm yields

$$h(z) = \exp\left(\sum_{k\geq 1}\left(\frac{z^k}{k} - \frac{z^{2k}}{2k^2} + \frac{z^{3k}}{3k^3} - \cdots\right)\right). \tag{4.25}$$

For $z < 1$, the series converges absolutely, permitting us to rearrange it as necessary. Our strategy will be to split the larger terms off from the beginning of the series, and sum them separately. First we have

$$\begin{aligned} h(z) &= \exp\left(\sum_{k\geq 1}\frac{z^k}{k} + \sum_{k\geq 1}\left(-\frac{z^{2k}}{2k^2} + \frac{z^{3k}}{3k^3} - \cdots\right)\right) \\ &= \frac{1}{1-z}g(z), \end{aligned} \tag{4.26}$$

where

$$g(z) = \exp\left(\sum_{k\geq 1}\left(-\frac{z^{2k}}{2k^2} + \frac{z^{3k}}{3k^3} - \cdots\right)\right). \tag{4.27}$$

In this form, h_n is the partial sum of the g_j coefficients in $g(z)$:

$$h_n = \sum_{0\leq j\leq n} g_j. \tag{4.28}$$

We will see later that the Tauberian limit theorem applies, hence

$$\begin{aligned} \lim_{n\to\infty} h_n &= \lim_{z\to 1-} g(z) \\ &= \exp\left(\sum_{k\geq 1}\left(-\frac{1}{2k^2} + \frac{1}{3k^3} - \cdots\right)\right) \\ &= \exp\left(\sum_{k\geq 1}\left(\ln\left(1+\frac{1}{k}\right) - \frac{1}{k}\right)\right) \\ &= \exp\left(\lim_{n\to\infty}\left(\ln(n+1) - H_n\right)\right) \\ &= e^{-\gamma}. \end{aligned} \tag{4.29}$$

Euler's constant, γ, appears mysteriously from the asymptotics for H_n, the harmonic numbers:

$$H_n = \ln n + \gamma + \frac{1}{2n} + O(n^{-2}). \tag{4.30}$$

Unfortunately part of the mystery lies in how fast h_n converges to this strange constant $e^{-\gamma}$. For error bounds, the Tauberian limit theorem is not particularly helpful. We must split another term off of the series in equation (4.25), and continue with a more detailed analysis:

$$g(z) = p(z^2)\, q(z) \tag{4.31}$$

where

$$\begin{aligned} p(z) &= \exp\left(-\frac{1}{2}\sum_{k\geq 1}\frac{z^k}{k^2}\right) \\ q(z) &= \exp\left(\sum_{k\geq 1}\left(\frac{z^{3k}}{3k^3} - \frac{z^{4k}}{4k^4} + \cdots\right)\right). \end{aligned} \tag{4.32}$$

First we attack $p(z)$ by deriving a recurrence relation for its coefficients:

$$p'(z) = p(z)\left(-\frac{1}{2}\sum_{k\geq 1}\frac{z^{k-1}}{k}\right) \tag{4.33}$$

$$-2n\, p_n = \sum_{0\leq k<n}\frac{p_k}{n-k}. \tag{4.34}$$

With this implicit formula we can use bootstrapping to derive a good estimate for p_n. To "prime the pump," it is easy to verify inductively that $p_n = O(1)$. Using this crude estimate in equation (4.34),

$$-2n\, p_n = \sum_{0\leq k<n}\frac{O(1)}{n-k}, \tag{4.35}$$

and replacing the right side with the asymptotics for the harmonic numbers, $O(\log n)$, gives an improved estimate of p_n:

$$p_n = O\left(\frac{\log n}{n}\right). \tag{4.36}$$

A further iteration of bootstrapping yields

$$p_n = O\left(\frac{\log n}{n}\right)^2. \tag{4.37}$$

At this point our estimate of p_n is good enough to begin dissecting the sum in equation (4.34). We wish to introduce more than a O-term in the asymptotics for p_n, so we remove the dominant part of the series in a form that is easy to sum:

$$\begin{aligned} -2np_n &= \sum_{0\le k<n} \frac{p_k}{n} + \sum_{0\le k<n} p_k \left(\frac{1}{n-k} - \frac{1}{n}\right) \\ &= \frac{1}{n}\sum_{k\ge 0} p_k - \frac{1}{n}\sum_{k\ge n} p_k + \frac{1}{n}\sum_{0\le k<n} p_k \left(\frac{k}{n-k}\right) \\ &= \frac{1}{n}p(1) - \frac{1}{n}\sum_{k\ge n} O\left(\frac{\log k}{k}\right)^2 + \frac{1}{n}\sum_{0\le k<n} O\left(\frac{(\log k)^2}{k(n-k)}\right) \\ &= \frac{1}{n}e^{-\pi^2/12} + O\left(\frac{(\log n)^3}{n^2}\right). \end{aligned} \tag{4.38}$$

In the last step we computed $p(1)$ by summing the infinite series

$$\sum_{k\ge 1} \frac{1}{k^2} = \zeta(2) = \frac{\pi^2}{6}. \tag{4.39}$$

We estimated the sum $\sum_{k\ge n} O\left(\frac{\log k}{k}\right)^2$ by considering its integral counterpart

$$\int_n^\infty \left(\frac{\log x}{x}\right)^2 dx = O\left(\frac{(\log n)^2}{n}\right). \tag{4.40}$$

And we estimated the remaining sum by computing with partial fractions:

$$\begin{aligned} \sum_{0\le k<n} O\left(\frac{(\log k)^2}{k(n-k)}\right) &= O\left((\log n)^2 \sum \frac{1}{k(n-k)}\right) \\ &= O\left(\frac{(\log n)^2}{n} \sum \left(\frac{1}{k} + \frac{1}{n-k}\right)\right) \\ &= O\left(\frac{(\log n)^3}{n}\right). \end{aligned} \tag{4.41}$$

Returning to equation (4.38), we now have a refined estimate of p_n,

$$p_n = \frac{-e^{-\pi^2/12}}{2n^2} + O\left(\frac{\log n}{n}\right)^3. \tag{4.42}$$

This expression can be bootstrapped through another iteration to obtain the slightly better approximation

$$p_n = \frac{-e^{-\pi^2/12}}{2n^2} + O\left(\frac{\log n}{n^3}\right). \tag{4.43}$$

Now that $p(z)$ is well understood, we turn our attention to the $q(z)$ portion remaining in equation (4.31). This time we split away the terms with $k = 1$, so that

$$q(z) = s(z)\, r(z) \tag{4.44}$$

where

$$\begin{aligned} s(z) &= \exp\left(\frac{z^3}{3} - \frac{z^4}{4} + \frac{z^5}{5} \cdots\right) \\ r(z) &= \exp\left(\sum_{k\geq 2}\left(\frac{z^{3k}}{3k^3} - \frac{z^{4k}}{4k^4} + \cdots\right)\right). \end{aligned} \tag{4.45}$$

The expression for $s(z)$ can be reworked,

$$\begin{aligned} s(z) &= \exp\left(\ln(1+z) - z + \frac{z^2}{2}\right) \\ &= (1+z)e^{-z+z^2/2}. \end{aligned} \tag{4.46}$$

From this we conclude that the coefficients, s_n, are exponentially small.

In $r(z)$, we collect terms with similar powers:

$$r(z) = \exp\left(\sum_{k\geq 3} z^k \sum_{\substack{3\leq d\leq k/2 \\ d \text{ divides } k}} \frac{\pm 1}{d\,(k/d)^d}\right). \tag{4.47}$$

The inside sum is $O(k^{-3})$. (This follows from the example used in Section 4.1.3 to illustrate dissecting sums.) Differentiating the formula for $r(z)$ and equating coefficients gives a recurrence relation for r_n:

$$r'(z) = r(z) \sum_{k\geq 3} k\, z^{k-1} O(k^{-3}) \tag{4.48}$$

$$n r_n = \sum_{0\leq k<n} r_k O\left(\frac{1}{n-k}\right)^2. \tag{4.49}$$

This recurrence can be bootstrapped repeatedly to give the successive bounds $r_n = O(1)$, $r_n = O(n^{-1})$, $r_n = O(n^{-2})$, and $r_n = O(n^{-3})$.

We have shattered our original problem into numerous fragments, but we have been able to deal effectively with each piece. Now we can begin to assemble the final result.

The pieces $r(z)$ and $s(z)$ combine to form $q(z)$ with coefficients

$$q_n = \sum_{0 \le k \le n} r_k s_{n-k}. \tag{4.50}$$

This is a convolution of two series that are $O(n^{-3})$, so the result is also $O(n^{-3})$. (To see this, divide the range into two parts, $0 \le k \le n/2$ and $0 \le n-k \le n/2$. What requirements on $f(n)$ suffice to make the convolution of two series that are $O(f(n))$ also $O(f(n))$?)

Next $q(z)$ and $p(z)$ combine to form $g(z)$:

$$g(z) = p(z^2)\, q(z) \tag{4.51}$$

$$g_n = \sum_{2k+l=n} p_k q_l. \tag{4.52}$$

Then g_n is summed to obtain h_n:

$$h_n = \sum_{j \le n} g_j = \sum_{2k+l \le n} p_k q_l. \tag{4.53}$$

We already know that the series on the right side of equation (4.53), when extended to infinity, converges to $e^{-\gamma}$, so we focus our attention on the tail:

$$\begin{aligned} h_n &= e^{-\gamma} - \sum_{2k+l>n} p_k q_l \\ &= e^{-\gamma} - \left(\sum_{l \ge 0} q_l \left(\sum_{2k>n} p_k + \sum_{n-l<2k\le n} p_k \right) \right). \end{aligned} \tag{4.54}$$

Using our earlier result for p_k, we can estimate the two internal sums. First

$$\begin{aligned} \sum_{2k>n} p_k &= \sum_{2k>n} \frac{-p(1)}{2k^2} + \sum_{2k>n} O\left(\frac{\log k}{k^3}\right) \\ &= \frac{-p(1)}{n} + O\left(\frac{\log n}{n^2}\right). \end{aligned} \tag{4.55}$$

Here we have used $p(1)$ instead of $e^{-\pi^2/12}$. This will prove useful when $p(1)$ and $q(1)$ combine to give $e^{-\gamma}$. In the last step we applied Euler's summation formula to both sums.

The other sum in equation (4.54) can be bounded by splitting it into two ranges,

$$\begin{aligned}\sum_{l\geq 0} q_l \sum_{n-l<2k\leq n} p_k &= \sum_{0\leq l<n/2} q_l \sum_{n-l<2k\leq n} p_k + \sum_{l\geq n/2} q_l \sum_{n-l\leq 2k\leq n} p_k \\ &= O\Bigl(\sum_{0\leq l<n/2} q_l \cdot l \cdot |p_{n/4}| + \sum_{l\geq n/2} q_l \cdot |p(1)|\Bigr) \\ &= O(n^{-2}). \qquad (4.56)\end{aligned}$$

Now that we have bounded all parts of equation (4.54), we can finally compute h_n:

$$\begin{aligned}h_n &= e^{-\gamma} + \frac{p(1)q(1)}{n} + O\left(\frac{\log n}{n^2}\right) \\ &= e^{-\gamma} + \frac{e^{-\gamma}}{n} + O\left(\frac{\log n}{n^2}\right). \qquad (4.57)\end{aligned}$$

Similar but simpler methods show that $g_n = O(n^{-1})$, so that our earlier use of the Tauberian theorem was indeed justified.

4.2 Stieltjes Integration and Asymptotics

Integrals are useful tools in asymptotics since they can be used to approximate discrete sums, and it is helpful to understand how an integral interacts with O-notation. For this reason we shall study the Stieltjes integral. The following definition and its immediate consequences are developed in [Apostol 57]:

Definition.

1) Let f and g be real-valued functions on $[a, b]$.

2) Let P be a partition of $[a, b]$ into $a = x_0 < x_1 < \ldots < x_n = b$.

3) Define a sum,

$$S(P) = \sum_{0 \le k < n} f(t_k)\left(g(x_{k+1}) - g(x_k)\right), \qquad t_k \in [x_k, x_{k+1}] \tag{4.58}$$

4) Then A is the value of the Stieltjes integral $\int_a^b f(t)\,dg(t)$ if and only if for all $\epsilon > 0$ there exists a P_ϵ such that all refinements P of P_ϵ lead to sums near A, that is, $|S(P) - A| < \epsilon$.

Consequences.

1) The Stieltjes integral has at most one value.

2) The Stieltjes integral is linear in f and g.

3) Adjacent intervals can be combined, $\int_a^b + \int_b^c = \int_a^c$.

4) (Integration by parts.) If $\int_a^b f(t)\,dg(t)$ exists then $\int_a^b g(t)\,df(t)$ exists and the sum of these two integrals is $f(t)g(t)|_a^b$.

5) (Change of variables by a continuous nondecreasing function h.)

$$\int_a^b f(h(t))\,dg(h(t)) = \int_{h(a)}^{h(b)} f(t)\,dg(t). \tag{4.59}$$

6) If $\int_a^b f(t)\,dg(t)$ exists and $g'(t)$ is continuous on $[a, b]$, then

$$\int_a^b f(t)\,dg(t) = \int_a^b f(t)g'(t)\,dt. \tag{4.60}$$

7) If a and b are integers and f is continuous from the right at integer points then

$$\int_a^b f(t)\,dg(\lceil t \rceil) = \sum_{a \le k < b} f(k)\,\Delta g(k), \qquad \Delta g(k) = g(k+1) - g(k). \tag{4.61}$$

7′) If a and b are integers and f is continuous from the left at integer points, then

$$\int_a^b f(t)\,dg(\lfloor t\rfloor) = \sum_{a<k\leq b} f(k)\,\nabla g(k), \qquad \nabla g(k) = g(k) - g(k-1). \quad (4.62)$$

8) If a and b are integers and g is continuous from the left at integer points, then

$$\int_a^b f(\lfloor t\rfloor)\,dg(t) = \sum_{a\leq k<b} f(k)\,\Delta g(k). \quad (4.63)$$

9) (Derivative of the integral.)

$$\int_a^b f(t)\,d\int_a^t g(u)\,dh(u) = \int_a^b f(t)\,g(t)\,dh(t). \quad (4.64)$$

10) If $\int_a^b f(t)\,dg(t)$ exists then $\int f(t)\,dg(t)$ exists for all subintervals of $[a,b]$.

11) $\int_a^b f(t)\,dg(t)$ exists if f is continuous and g is of bounded variation.

By bounded variation we mean that $\int_a^b |dg(t)|$ exists. Intuitively this implies that the variation, $\sum |g(x_{k+1}) - g(x_k)|$, gets small as the partition P gets small. Continuity is not enough, since $f(t) = g(t) = \sqrt{t}\cos(1/t)$ has no Stieltjes integral in intervals that include 0.

12) (Summation by parts.) Combining consequences 4, 7, and 7′, we obtain a very useful formula when a and b are integers:

$$\sum_{a\leq k<b} f(k)\,\Delta g(k) = f(k)g(k)\big|_a^b - \sum_{a<k\leq b} g(k)\,\nabla f(k). \quad (4.65)$$

4.2.1 O-notation and Integrals

The basic properties of Stieltjes integration allow us to derive two theorems stipulating when O can be removed from an integral.

Theorem 1.

$$\int_a^b O(f(t))\, dg(t) = O\Big(\int_a^b f(t)\, dg(t)\Big) \tag{4.66}$$

if g is monotone increasing, f is positive, and both integrals exist.

Proof. Recall that $a(t) = O(f(t))$ means that there is a constant M such that $|a(t)| < Mf(t)$. Since $f(t)$ and $dg(t)$ are nonnegative by hypothesis, we can bound the integral by $\int_a^b Mf(t)\, dg(t)$ and move M outside to derive the theorem.

Theorem 2.

$$\int_a^b f(t)\, dO\big(g(t)\big) = O\big(f(a)g(a)\big)+O\big(f(b)g(b)\big)+O\Big(\int_a^b f(t)\, dg(t)\Big) \tag{4.67}$$

when f and g are monotone increasing positive functions and the integrals exist.

Proof. Let $b(t)$ be the function that is $O\big(g(t)\big)$. We can integrate by parts and obtain

$$\int_a^b f(t)\, db(t) = f(t)b(t)\big|_a^b - \int_a^b b(t)\, df(t).$$

Theorem 1 applies to the last integral, hence we have

$$\int_a^b f(t)\, dO\big(g(t)\big) = O\big(f(a)g(a)\big)+O\big(f(b)g(b)\big)+O\Big(\int_a^b g(t)\, df(t)\Big). \tag{4.68}$$

Integration by parts is used again to exchange f and g, completing the proof of Theorem 2.

4.2.2 Euler's Summation Formula

Stieltjes integration provides a theoretical framework for the approximation of sums by integrals. Suppose we wish to approximate the sum of $f(k)$. We can begin with consequence number 7,

$$\sum_{a \le k < b} f(k) = \int_a^b f(t) d\lceil t \rceil . \tag{4.69}$$

Using the linearity property, the right-hand side can be expanded to

$$\int_a^b f(t)\,dt - \int_a^b f(t)\,d\left(t - \lceil t \rceil + \tfrac{1}{2}\right) + \int_a^b f(t)\,d\left(+\tfrac{1}{2}\right). \tag{4.70}$$

The first integral is a rough approximation to the sum; the second integral will allow us to refine the approximation; and the third integral is zero. A new term of Euler's summation formula appears when we integrate the second term by parts:

$$\begin{aligned} \sum_a^b f(k) &= \int_a^b f(t)\,dt - f(t)\left(t - \lceil t \rceil + \frac{1}{2}\right)\Bigg|_a^b + \int_a^b \left(t - \lceil t \rceil + \frac{1}{2}\right) df(t) \\ &= \int_a^b f(t)\,dt - \frac{1}{2} f(t)\Bigg|_a^b + \int_a^b \left(t - \lceil t \rceil + \frac{1}{2}\right) df(t). \end{aligned} \tag{4.71}$$

On the interval $[n, n+1]$ the last integral can be rewritten to read

$$\int_n^{n+1} \left(t - n - \frac{1}{2}\right) df(t) = \int_n^{n+1} f'(t)\,d\left(\frac{(t-n)^2 - (t-n) + 1/6}{2}\right). \tag{4.72}$$

And we can iterate this process, integrating by parts, and exchanging the rôles of f and g in the new integral $\int g\,df$.

There are several requirements necessary for such an iteration to work properly, and if we explore these requirements the mystery of the constants $1/2$ and $1/6$ will be revealed. First of all, we assume that $f'(t)$ exists. In fact each iteration will require a higher derivative of $f(t)$. The second requirement enters when we "integrate" the factor $(t - n - 1/2)$ and obtain $((t-n)^2 - (t-n) + 1/6)/2$. This change is made on each interval $[n, n+1]$, and from these segments the whole range is assembled. It is fortunate that $\left((t-n)^2 - (t-n) + 1/6\right)/2$ has the same value at n and $n+1$, so that the assembled integral $\int_a^b f'(t)\,dg(t)$ has a continuous function in the position of $g(t)$. Any discontinuities in g would make significant and unwanted contributions to the Stieltjes integral. The constant '1/2' in $(t - n - 1/2)$ is responsible for the continuity, and the constant '1/6' will guarantee a similar continuity when we integrate the polynomial again in the next iteration. We have a family of polynomials, $B_n(t - \lfloor t \rfloor)$, with the continuity condition $B_n(0) = B_n(1)$ holding at the endpoints for $n > 1$, satisfying the derivative relation $B_n'(x) = nB_{n-1}(x)$. These two requirements are sufficient to determine the Bernoulli polynomials:

$$\begin{aligned} B_1(x) &= x - 1/2 \\ B_2(x) &= x^2 - x + 1/6 \\ B_3(x) &= x^3 - (3/2)x^2 + (1/2)x \\ B_n(x) &= \sum_k \binom{n}{k} B_k x^{n-k} \end{aligned} \tag{4.73}$$

The constants B_k in the sum are the Bernoulli numbers:

$$B_0 = 1, \quad B_1 = -1/2, \quad B_2 = 1/6, \quad B_3 = 0, \quad B_4 = -1/30 \tag{4.74}$$

And these coefficients appear in the final summation formula:

$$\begin{aligned} \sum_{a \le k < b} f(k) = \int_a^b f(t)\,dt + B_1 f(t)\Big|_a^b + \frac{B_2}{2!} f'(t)\Big|_a^b + \cdots + \frac{B_{2m}}{(2m)!} f^{(2m-1)}(t)\Big|_a^b \\ + \int_a^b \frac{B_{2m+1}(t - \lfloor t \rfloor)}{(2m+1)!} f^{(2m+1)}(t)\,dt. \end{aligned} \tag{4.75}$$

(Strictly speaking, the sum implicitly represented by dots here has alternating signs,

$$\frac{B_2}{2!} f'(t)\Big|_a^b - \frac{B_3}{3!} f''(t)\Big|_a^b + \frac{B_4}{4!} f'''(t)\Big|_a^b - \cdots + \frac{B_{2m}}{(2m)!} f^{(2m-1)}(t)\Big|_a^b ;$$

but these signs are immaterial because the odd-numbered coefficients B_3, B_5, B_7, ... are all zero. See [GKP; Section 6.5] for further discussion of Bernoulli numbers.)

4.2.3 An Example from Number Theory

Suppose we have an integer n chosen at random from the interval $[1, x]$. The average number of distinct prime factors of n is given by the formula

$$\frac{1}{x}\sum_{n\le x}\sum_{p\backslash n} 1 = \frac{1}{x}\sum_{p\le x}\left\lfloor\frac{x}{p}\right\rfloor . \tag{4.76}$$

(Hereafter p will denote a prime. The notation $p\backslash n$ means "p divides n.") Ignoring the slight aberration caused by the floor function, the quantity of interest in the formula above is the sum of the reciprocals of primes $\le x$. We turn now to this restricted problem, where we will make several uses of Stieltjes integration. Initially, we can express the sum as an integral:

$$\sum_{p\le x}\frac{1}{p} = \int_{1.5}^{x}\frac{1}{t}\,d\pi(t), \qquad \pi(t) = \sum_{p\le t} 1. \tag{4.77}$$

Here $\pi(t)$ is a step function that changes only at the primes. The function $L(x)$ given by

$$L(x) = \int_{1.5}^{x}\frac{dt}{\ln t} \tag{4.78}$$

is known to give a close approximation to $\pi(x)$:

$$\pi(x) = L(x) + O\left(x\,e^{-c\sqrt{\log x}}\right). \tag{4.79}$$

(This strong form of the prime number theorem is due to de la Vallée Poussin in the 19th century; cf. [Knuth 76a].) By using $L(x)$ for $\pi(x)$ and applying Theorem 2 to remove O from the integral, we obtain an asymptotic estimate:

$$\begin{aligned}\sum_{p\le x}\frac{1}{p} &= \int_{1.5}^{x}\frac{dt}{t\ln t} + \int_{1.5}^{x}\frac{1}{t}\,dO\left(t\,e^{-c\sqrt{\log t}}\right)\\ &= \ln\ln x + O(1).\end{aligned} \tag{4.80}$$

Although we have no analog of Euler's summation formula for sums over primes, there is a roundabout way of improving this estimate. Using reasoning similar to that used above we can compute further sums:

$$C_m(x) = \sum_{p\le x}\frac{(\ln p)^m}{p} = \frac{(\ln x)^m}{m} + O(1), \qquad m \ge 1. \tag{4.81}$$

Then by Consequence 9, our original sum can be expressed as

$$C_0(x) = \int_{1.5}^{x} \frac{d\pi(t)}{t} = \int_{1.5}^{x} \frac{1}{(\ln t)^m}\, d\int_{1.5}^{t} \frac{(\ln u)^m\, d\pi(u)}{u} = \int_{1.5}^{x} \frac{dC_m(t)}{(\ln t)^m}. \tag{4.82}$$

And the last integral submits to integration by parts,

$$\begin{aligned} C_0(x) &= \frac{C_m(t)}{(\ln t)^m}\bigg|_{1.5}^{x} + m\int_{1.5}^{x} \frac{C_m(t)\,dt}{t(\ln t)^{m+1}} \\ &= \frac{1}{m} + O\left((\ln x)^{-m}\right) + \int_{1.5}^{x} \frac{dt}{t\ln t} + m\int_{1.5}^{x} \frac{O(1)\,dt}{t(\ln t)^{m+1}} \\ &= \ln\ln x + M + O\left((\log x)^{-m}\right), \text{ for some constant } M. \end{aligned} \tag{4.83}$$

This analysis applies to all $m > 0$, so we have proved a rather strong result about the asymptotics of the sum of reciprocal primes. However, the strength of the result makes the exact value of M a tantalizing question.

We can evaluate M by making use of the Riemann zeta function and Möbius inversion. The zeta function is related to prime numbers by

$$\zeta(s) = \sum_{n\geq 1} \frac{1}{n^s} = \prod_p \left(\frac{1}{1-p^{-s}}\right) = \prod_p \left(1 + p^{-s} + p^{-2s} + \cdots\right), \quad s > 1. \tag{4.84}$$

Following Euler, we will find it useful to work with the logarithm of this equation,

$$\ln\zeta(s) = \sum_p \left(\frac{1}{p^s} + \frac{1}{2p^{2s}} + \cdots\right) = \Sigma(s) + \frac{1}{2}\Sigma(2s) + \frac{1}{3}\Sigma(3s) + \cdots, \tag{4.85}$$

where $\Sigma(s) = \sum_p p^{-s}$. We are interested in the partial sums of the divergent series $\Sigma(1)$, and we can get information about them by considering the convergent series $\Sigma(s)$ for $s > 1$.

The Möbius function, defined by

$$\mu(n) = \begin{cases} 1, & \text{if } n = 1; \\ 0, & \text{if } n \text{ has a squared factor;} \\ (-1)^k, & \text{if } n \text{ has } k \text{ distinct prime factors;} \end{cases} \tag{4.86}$$

will invert formulas such as (4.85) above. The common form of Möbius inversion is

$$g(n) = \sum_{d\backslash n} f(d) \qquad \leftrightarrow \qquad f(n) = \sum_{d\backslash n} \mu(d) g\left(\frac{n}{d}\right). \tag{4.87}$$

But for our purposes we need another formulation,

$$g(x) = \sum_{m=1}^{\infty} f(mx) \qquad \leftrightarrow \qquad f(x) = \sum_{n=1}^{\infty} \mu(n) g(nx). \tag{4.88}$$

This allows us to express $\Sigma(s)$ in terms of $\zeta(s)$,

$$\Sigma(s) = \sum_{n} \mu(n) \frac{\ln \zeta(ns)}{n}. \tag{4.89}$$

Since $\zeta(s) = 1 + O(2^{-s})$ this last sum converges quickly to $\Sigma(s)$; we have a rapid way to evaluate $\Sigma(s)$ that will prove useful later when we express M in terms of $\Sigma(s)$. (These properties of the zeta and Möbius functions can be found, for example, in [Hardy 79; pp. 233–259].)

Let us pause a moment to plot strategy. We are interested in $\Sigma(1)$, but the formula above is valid only for $s > 1$. We could look at $\Sigma(1+\epsilon)$ and let $\epsilon \to 0$,

$$\Sigma(1+\epsilon) = \ln \zeta(1+\epsilon) - \frac{1}{2}\Sigma(2+2\epsilon) - \frac{1}{3}\Sigma(3+3\epsilon) + \cdots . \tag{4.90}$$

Standard references like [Hardy 79] give $\epsilon^{-1} + O(1)$ for the asymptotics of $\zeta(1+\epsilon)$ near 1, so this simplifies to

$$\Sigma(1+\epsilon) = -\ln \epsilon - \sum_{n=2}^{\infty} \frac{\Sigma(n+n\epsilon)}{n} + O(\epsilon). \tag{4.91}$$

Unfortunately this formula blows up in a different sense than our original expression,

$$C_0(x) = \sum_{p \le x} \frac{1}{p} = \ln \ln x + M + O\left((\log x)^{-m}\right), \tag{4.92}$$

does. So we cannot simply cancel the leading terms of the two formulas to obtain information about M. Instead we must rework the C_0 formula to depend on ϵ.

To rework C_0, we introduce ϵ so that x can be sent to infinity,

$$\sum_{p} \frac{1}{p^{1+\epsilon}} = \int_{1.5}^{\infty} \frac{d\pi(t)}{t^{1+\epsilon}} = \lim_{x \to \infty} \int_{1.5}^{x} \frac{dC_0(t)}{t^{\epsilon}}; \tag{4.93}$$

here again we have used Consequence 9 to replace $d\pi(t)$ with $t\,dC_0(t)$. Integrating by parts gives

$$\Sigma(1+\epsilon) = \lim_{x\to\infty} \left(\frac{C_0(x)}{x^\epsilon} - \int_{1.5}^{x} C_0(t) d\left(t^{-\epsilon}\right) \right). \tag{4.94}$$

Now the old asymptotics for C_0 will replace C_0 in the integral. By these same asymptotics $C_0(x)/x^\epsilon$ vanishes. This leaves

$$\Sigma(1+\epsilon) = \epsilon \left(\int_{1.5}^{\infty} \left(\ln\ln t + M + O\left((\log t)^{-1}\right)\right) \frac{dt}{t^{1+\epsilon}} \right). \tag{4.95}$$

Next we substitute $e^{u/\epsilon}$ for t, obtaining

$$\Sigma(1+\epsilon) = \int_{\epsilon \ln 1.5}^{\infty} e^{-u} \left(\ln u - \ln \epsilon + M + O\left(\frac{\epsilon}{u}\right) \right) du. \tag{4.96}$$

Most terms of this integral are easy to deal with, except $e^{-u}\ln u$ which can be expressed in terms of the exponential integral:

$$\int_a^\infty e^{-u} \ln u \, du = e^{-a} \ln a + \int_a^\infty \frac{e^{-u}}{u} du. \tag{4.97}$$

For small a the exponential integral has well understood asymptotics,

$$E_1(a) = \int_a^\infty \frac{e^{-u}}{u} du = -\ln a - \gamma + O(a). \tag{4.98}$$

Applying our knowledge of $E_1(a)$ to equation (4.96) gives

$$\begin{aligned} \Sigma(1+\epsilon) &= (1.5)^{-\epsilon}(\ln\epsilon + \ln\ln 1.5) + E_1(\epsilon \ln 1.5) \\ &\quad - (1.5)^{-\epsilon} \ln \epsilon + (1.5)^{-\epsilon} M + O\left(\epsilon E_1(\epsilon \ln 1.5)\right) \\ &= -\ln\epsilon - \gamma + M + O\left(\epsilon \ln \frac{1}{\epsilon}\right). \end{aligned} \tag{4.99}$$

Now we can compare this reworked formula with our previous expression (4.91) for $\Sigma(1+\epsilon)$, to derive the desired formula for M:

$$M = \gamma - \frac{1}{2}\Sigma(2) - \frac{1}{3}\Sigma(3) - \cdots. \tag{4.100}$$

Since $\Sigma(s) = O\left(2^{-s}\right)$ this series converges rapidly; the precise value of M is 0.26149 72128 47643 ([Mertens 1874], [Knuth 76a]).

Returning to the question raised at the beginning of the section, we find that the average number of distinct prime factors of n can be computed from the results above:

$$\begin{aligned}\frac{1}{x}\sum_{p\leq x}\left\lfloor\frac{x}{p}\right\rfloor &= \frac{1}{x}\sum_{p\leq x}\left(\frac{x}{p}+O(1)\right)\\ &= \sum_{p\leq x}\frac{1}{p}+O\left(\frac{1}{\log x}\right)\\ &= \ln\ln x + M + O\left(\frac{1}{\log x}\right).\end{aligned}\tag{4.101}$$

4.3 Asymptotics from Generating Functions

Frequently a combinatorial argument will produce a generating function, $G(z)$, with interesting coefficients that have no simple closed form. This section will address two popular techniques for obtaining the asymptotics of these g_n for large n, given $G(z) = \sum g_n z^n$. The choice of technique depends on the nature of $G(z)$: If $G(z)$ has singularities, then a Darboux approach can use these singularities to obtain the asymptotics of g_n. On the other hand, if $G(z)$ converges everywhere we employ the saddle point method to find and evaluate a contour integral.

4.3.1 Darboux's Method

When $G(z)$ converges in a circle of radius R, the sum $\sum |g_n r^n|$ converges absolutely for $r < R$ and this is possible only if $g_n = O(r^{-n})$. This basic fact about series suggests the following (somewhat idealized) approach to the asymptotics of g_n. When $G(z)$ has a singularity at radius R we find a function $H(z)$ with well known coefficients that has the same singularity. Then $G(z) - H(z)$ will often have a greater radius of convergence, S, and g_n will be well approximated by h_n:

$$g_n = h_n + O\left(s^{-n}\right), \qquad s < S. \tag{4.102}$$

The process is repeated until S is extended far enough to provide a small error bound.

The method depends critically on finding a comparison function $H(z)$ with well known coefficients. If we are attempting to cancel an ordinary pole at $z = a$ in $G(z)$, then $H(z)$ is easy to construct since $G(z)$ will have the Laurent form

$$G(z) = \frac{C_{-m}}{(z-a)^m} + \cdots + \frac{C_{-1}}{(z-a)} + C_0 + C_1(z-a) + \cdots. \tag{4.103}$$

For $H(z)$ we use the terms with negative powers of $(z-a)$ in the expansion:

$$H(z) = \frac{C_{-m}}{(z-a)^m} + \frac{C_{-m+1}}{(z-a)^{m-1}} + \cdots + \frac{C_{-1}}{(z-a)}. \tag{4.104}$$

The coefficients of $H(z)$ can be obtained with the binomial expansion of

$$(z-a)^{-j} = (-a)^{-j} \sum_k \binom{-j}{k} \left(\frac{-z}{a}\right)^k. \tag{4.105}$$

(See [Knuth III; pp. 41–42] for an illustration of Darboux's technique applied to a function where the singularities are poles.)

Algebraic singularities are considerably harder to remove; in fact we will only be able to "improve" the singularity in a vague sense that will become clear shortly. By algebraic we mean that $G(z)$ can be expressed as a finite sum of terms of the form

$$(z-a)^{-w}g(z), \quad w \text{ complex}, \quad g(z) \text{ analytic at } a. \tag{4.106}$$

For example,

$$\sqrt{1-z} = \sum_n \binom{1/2}{n}(-z)^n \tag{4.107}$$

has an algebraic singularity at $z = 1$, although in this case the function also has a binomial expansion so that Darboux's method is unnecessary. Darboux's technique will be illustrated with the function

$$G(z) = \sqrt{(1-z)(1-\alpha z)}, \qquad \alpha < 1. \tag{4.108}$$

(See [Knuth I; exercise 2.2.1–12].) We need a comparison function that will attack the singularity at $z = 1$, so we first expand

$$\sqrt{1-\alpha z} = \sqrt{1-\alpha} + C_1(1-z) + C_2(1-z)^2 + \cdots. \tag{4.109}$$

The first term of the expansion suggests choosing the comparison function

$$H(z) = \sqrt{1-z}\,\sqrt{1-\alpha}\,; \tag{4.110}$$

further terms of the expansion can be used to improve the estimate. Let us see how well $H(z)$ performs by itself:

$$\begin{aligned} G(z) - H(z) &= \sqrt{1-z}\,\left(\sqrt{1-\alpha z} - \sqrt{1-\alpha}\right) \\ &= \alpha(1-z)^{3/2}\left(\frac{1}{\sqrt{1-\alpha z}+\sqrt{1-\alpha}}\right) \\ &= A(z)\,B(z) \end{aligned} \tag{4.111}$$

where

$$\begin{aligned} A(z) &= \alpha(1-z)^{3/2} \\ B(z) &= 1/\left(\sqrt{1-\alpha z}+\sqrt{1-\alpha}\right). \end{aligned} \tag{4.112}$$

Note that we have not removed the singularity at $z = 1$, but instead we have "improved" the singularity from $(1-z)^{1/2}$ to $(1-z)^{3/2}$. This improvement is strong enough to make $H(z)$ a good approximation to $G(z)$.

The error is the coefficient of z^n in $A(z)B(z)$. The power series $B(z)$ has a radius of convergence greater than 1, and so $b_n = O(r^{-n})$ for some $r > 1$. Furthermore $A(z)$ can be expanded,

$$A(z) = \alpha \sum_{n\geq 0} \binom{3/2}{n} (-z)^n = \alpha \sum_{n \geq 0} \binom{n-5/2}{n} z^n, \tag{4.113}$$

and this gives $a_n = \alpha\binom{n-5/2}{n} = O\left(n^{-5/2}\right)$. To derive the error bound we proceed as in Section 4.1 to split the convolution of $A(z)$ and $B(z)$ into two sums:

$$\begin{aligned} \sum_{0\leq k \leq n/2} a_k b_{n-k} &= O\left(r^{-n/2}\right) \\ \sum_{n/2 < k \leq n} a_k b_{n-k} &= O\left(n^{-5/2}\right). \end{aligned} \tag{4.114}$$

Thus we may assert that

$$g_n = \sqrt{1-\alpha}\,(-1)^n \binom{1/2}{n} + O\left(n^{-5/2}\right). \tag{4.115}$$

In retrospect, our derivation of g_n is simply an expansion of $G(z)$ about $z = 1$. The error term is tricky, but depends on increasing the exponent of $(1-z)$ from $1/2$ to $3/2$. In fact a similar exponent dependency appears in the statement of Darboux's theorem below. The notion of weight is introduced and we "improve" the singularities by decreasing their weight:

Theorem. *Suppose $G(z) = \sum_{n\geq 0} g_n z^n$ is analytic near 0 and has only algebraic singularities on its circle of convergence. The singularities, resembling*

$$(1 - z/\alpha)^{-w} h(z), \tag{4.116}$$

are given weights equal to the real parts of their w's. Let W be the maximum of all weights at these singularities. Denote by α_k, w_k, and $h_k(z)$ the values of α, w, and $h(z)$ for those terms of the form (4.116) of weight W. Then

$$g_n = \frac{1}{n} \sum_k \frac{h_k(\alpha_k) n^{w_k}}{\Gamma(w_k)\alpha_k^n} + o\left(s^{-n} n^{W-1}\right), \tag{4.117}$$

where $s = |\alpha_k|$, the radius of convergence of $G(z)$, and $\Gamma(z)$ is the Gamma function.

This version of Darboux's theorem, found in [Bender 74], gives the first term of the asymptotics by diminishing all the heavy weight singularities.

The process can be repeated, resulting in the slightly more complicated statement of the theorem found in [Comtet 74]. An ordinary pole corresponds to integer w in the theorem, in which case repeated application will eventually reduce w to 0, eliminating the singularity completely, since the values $w = 0, -1, -2, \ldots$ are not singularities. The elementary method we have illustrated in our analysis of (4.108) is powerful enough to prove Darboux's general theorem [Knuth 89].

4.3.2 Residue Calculus

The residue theorem states that the integral around a closed curve in the complex plane can be computed from the residues at the enclosed poles:

$$\frac{1}{2\pi i}\oint_C f(z)\,dz = \sum(\text{residues at enclosed poles}). \tag{4.118}$$

Here the residue of $f(z)$ at a is defined to be the coefficient C_{-1} in the Laurent expansion of $f(z)$:

$$f(z) = \frac{C_{-m}}{(z-a)^m} + \cdots + \frac{C_{-1}}{(z-a)} + C_0 + C_1(z-a) + \cdots. \tag{4.119}$$

Residues are relatively easy to compute. If $m = 1$ then the pole is first order and the residue is given by

$$C_{-1} = \lim_{z\to a}(z-a)f(z). \tag{4.120}$$

The limit usually succumbs to repeated application of l'Hospital's rule:

$$\lim_{z\to a}\frac{g(z)}{h(z)} = \lim_{z\to a}\frac{g'(z)}{h'(z)}. \tag{4.121}$$

A pole is considered to be of order m if $\lim_{z\to a}(z-a)^m f(z)$ is a nonzero constant, but this limit does not tell us anything about the residue when $m > 1$. Differentiation can be used to isolate the correct coefficient:

$$C_{-1} = \frac{1}{(m-1)!}\lim_{z\to a}\frac{d^{m-1}}{dz^{m-1}}\left((z-a)^m f(z)\right). \tag{4.122}$$

In practice, however, it is often faster to deduce the behavior of $f(z)$ near $z = a$ by substituting $z = a + w$ and expanding in powers of w, then to obtain the coefficient of w^{-1} by inspection.

Traditionally the residue theorem is given as an easy way to compute the integral on a closed curve. In asymptotics we often use the formula backwards, placing the combinatorial quantity of interest in the residue and then evaluating the integral.

For example, suppose we have a double generating function,

$$F(w,z) = \sum_{m,n} a_{mn} w^m z^n, \tag{4.123}$$

and we wish to compute a generating function for the diagonal elements,

$$G(z) = \sum_{n} a_{nn} z^n. \tag{4.124}$$

Terms with $n = m$ are moved to the coefficient of t^{-1}, where they become the residue:

$$\begin{aligned}\frac{1}{2\pi i}\oint F(t, z/t)\,\frac{dt}{t} &= \frac{1}{2\pi i}\oint \left(\sum_{m,n} a_{mn} t^m \left(\frac{z}{t}\right)^n\right)\frac{dt}{t}\\ &= \frac{1}{2\pi i}\sum_{m,n}\oint a_{mn} t^{m-n} z^n \frac{dt}{t}\\ &= \sum_n a_{nn} z^n = G(z).\end{aligned} \tag{4.125}$$

This interchange of summation and integration is legitimate if the series converges uniformly, so the path of integration must be chosen to make both $|t|$ and $|z/t|$ sufficiently small.

A classic illustration of the diagonalization of power series begins with

$$F(w,z) = \sum_{m,n\geq 0}\binom{m+n}{n} w^m z^n = \frac{1}{1-w-z}. \tag{4.126}$$

We seek an expression for the generating function

$$G(z) = \sum_n \binom{2n}{n} z^n. \tag{4.127}$$

Using the formula derived above,

$$G(z) = \frac{1}{2\pi i}\oint_C \frac{dt}{(1-t-z/t)t}. \tag{4.128}$$

If C is chosen to be a small curve around the origin, it encloses the first order pole at

$$t = \frac{1-\sqrt{1-4z}}{2}. \tag{4.129}$$

Here the residue is $(1-4z)^{-1/2}$ so the value of the integral is

$$G(z) = \frac{1}{\sqrt{1-4z}}. \tag{4.130}$$

For a second illustration of diagonalization consider the problem of obtaining the termwise product of two power series,

$$A(z) = \sum a_n z^n, \quad \text{and} \quad B(z) = \sum b_n z^n. \tag{4.131}$$

Using the result derived in (4.125), we obtain the Hadamard product:

$$G(z) = \sum a_n b_n z^n = \frac{1}{2\pi i} \oint A(t) B\left(\frac{z}{t}\right) \frac{dt}{t}. \tag{4.132}$$

4.3.3 The Saddle Point Method

Our next example makes use of several standard techniques that deserve attention before we begin the actual problem. Initially, we will use the residue theorem backwards:

$$g_n = \frac{1}{2\pi i} \oint \frac{G(z)dz}{z^{n+1}}. \tag{4.133}$$

A generating function $G(z)$ is given, and we assume that it is free of singularities (otherwise a Darboux attack would provide the asymptotics) so that the only constraint on the path of integration is that it encloses the origin. A wise choice for this path allows the integral to be easily estimated, and a good heuristic for choosing paths is the saddle point method. The idea is to run the path of integration through a saddle point, which is defined to be a place where the derivative of the integrand is zero. Like a lazy hiker, the path then crosses the ridge at a low point; but unlike the hiker, the best path takes the steepest ascent to the ridge. In fact, for our purposes, this property is far more important than crossing the ridge at the lowest point.

Once we have chosen a path of integration another technique, Laplace's method for integrals, is frequently helpful. The integral will be concentrated in a small interval, but will include negligible tails extending over the whole region. Laplace's method removes these tails and replaces them by a different small function that is convenient for the evaluation of the integral. Both the old tails and the new tails must be shown to be insignificant to the result of the evaluation.

As an example problem, we will derive a strong version of the central limit theorem, which states that the mean of a large number of drawings from an arbitrary distribution is normally distributed:

$$\text{Prob}\left(\mu - \frac{\alpha\sigma}{\sqrt{n}} < \frac{X_1 + X_2 + \cdots + X_n}{n} < \mu + \frac{\beta\sigma}{\sqrt{n}}\right) = \frac{1}{\sqrt{2\pi}} \int_{-\alpha}^{\beta} e^{-z^2/2} dz \left(1 + O(n^{-1})\right). \quad (4.134)$$

Here the X_i are arbitrary but identically distributed random variables with mean μ and standard deviation σ. With several minor restrictions we can prove an even stronger result, clarifying exactly how fast an arbitrary integer-valued random variable converges to the normal distribution.

Assumption 1. The X_i are drawn from an integer-valued distribution with generating function $g(z)$:

$$g(z) = \sum_{k \geq 0} p_k z^k, \qquad p_k = \text{Prob}(X_i = k). \quad (4.135)$$

We assume that $g(z)$ is analytic for $|z| < 1 + \delta$, and since $g(1) = 1$ for a probability distribution, we may conclude that $\ln g(e^t)$ is analytic at $t = 0$. This allows us to characterize $g(z)$ by its Thiele expansion,

$$g(e^t) = \exp\left(\mu t + \frac{\sigma^2 t^2}{2!} + \frac{\kappa_3 t^3}{3!} + \frac{\kappa_4 t^4}{4!} + \cdots\right). \quad (4.136)$$

where κ_j is the jth semi-invariant of $g(z)$.

Assumption 2. $g(0)$ must be nonzero, that is $p_0 \neq 0$. This is not a restriction, since we can translate the generating function to $z^{-m}g(z)$, where p_m is the first nonzero coefficient.

Assumption 3. The greatest common divisor of all k with $p_k \neq 0$ must be 1. This is also not a restriction since we may analyze $g(z^{1/m})$, where m is the greatest common divisor of the k such that $p_k \neq 0$.

The sum of n drawings from $g(z)$ has distribution $g(z)^n$. We wish to understand the behavior of this sum near its mean, μn, so we define

$$A_{n,r} = \text{coefficient of } z^{\mu n+r} \text{ in } g(z)^n, \tag{4.137}$$

where r is chosen to make $\mu n + r$ an integer. By the residue theorem,

$$A_{n,r} = \frac{1}{2\pi i} \oint \frac{g(z)^n dz}{z^{\mu n+r+1}}. \tag{4.138}$$

The saddle point is near $z = 1$, so we choose a path of integration with radius 1 enclosing the origin, and substitute $z = e^{it}$:

$$A_{n,r} = \frac{1}{2\pi} \int_{-\pi}^{\pi} \frac{g(e^{it})^n dt}{e^{it(\mu n+r)}}. \tag{4.139}$$

Assumption 3 implies that the terms of $g(e^{it})$ will all be in phase only when $t = 0$, so $|g(e^{it})| < 1$, except when $t = 0$, in which case $g(1) = 1$. Raising $g(e^{it})$ to the nth power makes the tails of the integral exponentially small, and leaves the primary contribution at $t = 0$. In particular, whenever we choose a $\delta > 0$ there exists an $\alpha \in [0,1)$ such that $|g(e^{it})| < \alpha$ for $\delta \le |t| \le \pi$, and this means that

$$A_{n,r} = \frac{1}{2\pi} \int_{-\delta}^{\delta} \frac{g(e^{it})^n dt}{e^{it(\mu n+r)}} + O(\alpha^n). \tag{4.140}$$

Laplace's technique suggests that we chop off the tails and replace them with more agreeable functions. We will make three passes at the present tails, refining the interval each time, before adding new tails. First we set δ_1 small enough so that the Thiele expansion for $g(e^t)$ is valid, and then expand

$$\frac{g(e^{it})^n}{e^{it(\mu n+r)}} = \exp\left(-irt - \frac{\sigma^2 t^2 n}{2!} - \frac{i\kappa_3 t^3 n}{3!} + \frac{\kappa_4 t^4 n}{4!} + \cdots\right). \tag{4.141}$$

Next we set δ_2 smaller than δ_1 so that the first two terms in the expansion dominate the remaining terms:

$$\left| -\frac{i\kappa_3 t^3 n}{3!} + \frac{\kappa_4 t^4 n}{4!} + \cdots \right| < \frac{\sigma^2 \left|t^2\right| n}{6}, \qquad \text{for } |t| < \delta_2. \tag{4.142}$$

These first two refinements permit a third refinement from $[-\delta_2, \delta_2]$ to $[-n^{-1/2+\epsilon}, n^{-1/2+\epsilon}]$. (The role of the mysterious epsilon will become apparent shortly.) The error introduced by this refinement is the sum of two terms like

$$\int_{n^{-1/2+\epsilon}}^{\delta_2} \left|\exp\left(-irt - \frac{\sigma^2 t^2 n}{2!}\right)\right| \cdot \left|\exp\left(-\frac{i\kappa_3 t^3 n}{3!} + \frac{\kappa_4 t^4 n}{4!} + \cdots\right)\right| dt. \tag{4.143}$$

The $-irt$ in the first term contributes an irrelevant phase, and the second term is bounded by equation (4.142), so the error is exponentially small:

$$\int_{n^{-1/2+\epsilon}}^{\delta_2} \exp\left(-\frac{\sigma^2 t^2 n}{2} + \frac{\sigma^2 t^2 n}{6}\right) dt \le \delta_2 e^{-\sigma^2 n^{2\epsilon}/3}. \tag{4.144}$$

The reason for choosing $n^{-1/2+\epsilon}$ should now be clear. In the last step we bounded the integral by its largest value, substituting $n^{-1/2+\epsilon}$ for t in the integrand. The $n^{-1/2}$ exactly cancels the n associated with t^2 in the integrand, so ϵ becomes the "straw that breaks the camel's back" and drives the integral to zero.

We can summarize the progress so far by claiming that there exists an $\alpha \in (0,1)$ such that, for all $\epsilon > 0$,

$$A_{n,r} = \frac{1}{2\pi} \int_{-n^{-1/2+\epsilon}}^{n^{-1/2+\epsilon}} \frac{g(e^{it})^n\, dt}{e^{it(n\mu+r)}} + O\left(\alpha^{n^{2\epsilon}}\right). \tag{4.145}$$

Within such a small interval, the first two terms of the Thiele expansion are of principal importance:

$$A_{n,r} = \frac{1}{2\pi} \int_{-n^{-1/2+\epsilon}}^{n^{-1/2+\epsilon}} \exp\left(-irt - \frac{\sigma^2 t^2 n}{2} + O\left(n^{3\epsilon-1/2}\right)\right) dt + O\left(\alpha^{n^{2\epsilon}}\right). \tag{4.146}$$

At this point we are ready to add new tails to the integral, using the first two terms of the Thiele expansion as a convenient function. The new tails are exponentially small,

$$\left|\int_{n^{-1/2+\epsilon}}^{\infty} \exp\left(-irt - \frac{\sigma^2 t^2 n}{2}\right) dt\right| \le \int_{n^{-1/2+\epsilon}}^{\infty} \exp\left(\frac{-\sigma^2 t^2 n}{2}\right) dt = O\left(\exp\left(\frac{-\sigma^2 n^{2\epsilon}}{2}\right)\right).$$

It is then an easy matter to evaluate the integral on $[-\infty, \infty]$ by completing the square in the exponent:

$$A_{n,r} = \frac{1}{2\pi}\int_{-\infty}^{\infty} \exp\left(-irt - \frac{\sigma^2 t^2 n}{2}\right)\left(1 + O\left(n^{3\epsilon - 1/2}\right)\right) dt + O\left(\alpha^{n^{2\epsilon}}\right)$$
$$= \frac{1}{\sigma\sqrt{2\pi n}} \exp\left(\frac{-r^2}{2\sigma^2 n}\right) + O\left(n^{3\epsilon-1}\right). \tag{4.147}$$

Note that the constant implied by this O depends on $g(t)$ and ϵ but not on n or r.

We can improve this result by using more of the Thiele expansion in equation (4.146). This will require integrating terms like

$$\int_{-\infty}^{\infty} e^{iat - bt^2} t^k dt. \tag{4.148}$$

Completing the square in the exponent and expanding t^k with the binomial theorem will lead to still more terms of the form

$$\int_{-\infty}^{\infty} e^{-u^2} u^j du. \tag{4.149}$$

For odd j the integral vanishes, and for even j it can be transformed to the Gamma function by the substitution $v = u^2$. With this machinery we can extend our estimate, obtaining for example

$$A_{n,r} = \frac{1}{\sigma\sqrt{2\pi n}} \exp\left(\frac{-r^2}{2\sigma^2 n}\right)\left(1 - \frac{\kappa_3}{2\sigma^4}\left(\frac{r}{n}\right) + \frac{\kappa_3}{6\sigma^6}\left(\frac{r^3}{n^2}\right)\right) + O(n^{-3/2}). \tag{4.150}$$

The coefficient of the general term in the expansion, r^R/n^N, is given by

$$\sum_{S\geq 0} \frac{(-1)^S (R+2S)^{\underline{2S}}}{\sigma^{2(R+S)} 2^S S!} \sum_{\substack{p_3+p_4+p_5+\cdots = R-N+S \\ 3p_3+4p_4+5p_5+\cdots = R+2S}} \frac{1}{p_3!}\left(\frac{\kappa_3}{3!}\right)^{p_3} \frac{1}{p_4!}\left(\frac{\kappa_4}{4!}\right)^{p_4} \cdots; \tag{4.151}$$

such terms are present for $0 \leq R \leq \frac{3}{2}N$.

This is a very strong result; the central limit theorem follows immediately by summing on r, and if necessary, we have a detailed understanding of the asymptotic behavior of individual terms of the distribution. Unfortunately the formula above suffers from a weakness that is common to central limit theorems: Its range is limited. Note that since the error term is polynomial in n, the estimate of $A_{n,r}$ is useful only when $r = O(\sqrt{n})$. This is not surprising, since we were sloppy in our choice of the path of integration; it goes through the lowest portion of the saddle point only when $r = 0$, and becomes progressively worse for larger r. The obvious remedy would be to change the path of integration when r exceeds $\sqrt{n}$. However, we will see in a moment that the distribution itself can be shifted. This often proves to be easier than repeating the derivation with a different path of integration, although both techniques are essentially the same.

The coefficient of z^m in $g(z)^n$ can be obtained with the formula

$$[z^m]\,(g(z)^n) = \frac{g(\alpha)^n}{\alpha^m}\,[z^m]\left(\frac{g(\alpha z)}{g(\alpha)}\right)^n. \tag{4.152}$$

The right side of the equation seems like an unnecessary complication of the left side, since both require extracting the coefficient of z^m; but in fact the right side has an extra degree of freedom represented by α, which allows us to shift the mean of the distribution to a value close to m/n. We do this by choosing α so that

$$\frac{\alpha g'(\alpha)}{g(\alpha)} = \frac{m}{n}. \tag{4.153}$$

Take, for a simple example, the problem of finding the coefficient of $z^{n/3}$ in the binomial distribution with parameter $1/2$,

$$[z^{n/3}]\left(\frac{1+z}{2}\right)^n. \tag{4.154}$$

The coefficient of interest is at a distance $\gg \sqrt{n}$ from the mean, $n/2$, so equation (4.150) is useless until we shift the mean to $n/3$ by appropriate choice of α:

$$\frac{\alpha g'(\alpha)}{g(\alpha)} = \frac{\alpha}{1+\alpha} = \frac{1}{3}, \tag{4.155}$$

$$\alpha = \frac{1}{2}. \tag{4.156}$$

The new distribution,

$$\left(\tfrac{2}{3} + \tfrac{1}{3}z\right), \tag{4.157}$$

has mean $\mu = 1/3$, standard deviation $\sigma = \sqrt{2}/3$, $\kappa_3 = 2/27$, and $\kappa_4 = -2/27$. We apply (4.150) and (4.151) with $r = 0$, obtaining

$$\begin{aligned}[z^{n/3}]\left(\tfrac{2}{3} + \tfrac{1}{3}z\right)^n &= A_{n,0}\\ &= \frac{3}{2\sqrt{\pi n}}\left(1 - \frac{7}{24n}\right) + O(n^{-5/2}).\end{aligned} \tag{4.158}$$

Multiplying by the $g(\alpha)^n/\alpha^m$ factor found in equation (4.152) gives a solution to the original problem of estimating the probability of exactly $n/3$ heads appearing in n tosses of a fair coin:

$$2^{n/3}\left(\frac{3}{4}\right)^n \frac{3}{2\sqrt{\pi n}}\left(1-\frac{7}{24n}+O(n^{-2})\right). \tag{4.159}$$

Lest the reader be left with the impression that shifting the mean is a panacea for all range problems, several difficulties should be mentioned. In equation (4.155) we were fortunate to find α constant. In general α will have some dependency on n, which in turn will make the mean, standard deviation, and other semi-invariants dependent on n. Our derivation of a strong version of the central limit theorem made no allowance for this dependency, and must be reworked to accommodate the specific problem. In particular, the application of Laplace's method (shaving the tails of the integral and adding new tails) is likely to be affected by the new variations. Nevertheless, shifting the mean is still useful as a clear guide for the asymptotic derivation, and the reader will find it interesting to derive asymptotic formulas for Stirling numbers in this way.

Bibliography

[Aho 73] Aho, A. V. and Sloane, N. J. A.
Some Doubly Exponential Sequences
Fibonacci Quarterly 11(4):429–437, 1973

[Amble 74] Amble, O. and Knuth, D. E.
Ordered Hash Tables
The Computer Journal 17 (2):135–142, 1974

[Apostol 57] Apostol, T.
Mathematical Analysis:
A Modern Approach to Advanced Calculus
Addison-Wesley, 1957

[Bailey 35] Bailey, W. N.
Generalized Hypergeometric Series
Cambridge Tracts in Mathematics and Mathematical Physics, No. 32; Cambridge University Press, 1935

[Bender 74] Bender, E. A.
Asymptotic Methods in Enumeration
SIAM Review 16 (4):485–515, 1974

[Boyce 69] Boyce, W. and DiPrima, R.
Elementary Differential Equations
and Boundary Value Problems
John Wiley and Sons, 1969

[Comtet 74] Comtet, L.
Advanced Combinatorics
D. Reidel Publishing Co., 1974

[deBruijn 70] de Bruijn, N. G.
Asymptotic Methods in Analysis
North-Holland Publishing Co., 1970

[deBruijn 72] de Bruijn, N. G., Knuth, D. E., and Rice, S. O.
The Average Height of Planted Plane Trees
In R. C. Read, editor, *Graph Theory and Computing*:15–22
Academic Press, 1972

[Delange 75] Delange, H.
Sur la fonction sommatoire de la fonction somme des chiffres
Enseignement Mathématique 21 (1):31–47, 1975

[Erdős 59] Erdős, P. and Rényi, A.
On Random Graphs I
Publicationes Mathematicæ 6:385–387, 1959

[Erdős 60] Erdős, P. and Rényi, A.
On the Evolution of Random Graphs
Magyar Tudományos Akadémia:
Matematikai Kutató Intézetének Közleményei 5(1):17–61, 1960

[Fredman 74] Fredman, M. L., and Knuth, D. E.
Recurrence Relations Based on Minimization
Journal of Mathematical Analysis and Applications
48(2):534–559, 1974

[Gould 73] Gould, H. and Hsu, L.
Some New Inverse Relations
Duke Mathematical Journal 40(4):885–892, 1973

[GKP] Graham, R. L., Knuth, D. E., and Patashnik, O.
Concrete Mathematics
Addison-Wesley, 1988

[Hardy 49] Hardy, G. H.
Divergent Series
Oxford, 1949

[Hardy 79] Hardy, G. H., and Wright, E. M.
An Introduction to the Theory of Numbers
Oxford, 1979

[Henrici I] Henrici, P.
Applied and Computational Complex Analysis, Volume 1
John Wiley and Sons, 1974

[Henrici II] Henrici, P.
Applied and Computational Complex Analysis, Volume 2
John Wiley and Sons, 1977

[Jonassen 78] Jonassen, A., and Knuth, D. E.
A Trivial Algorithm Whose Analysis Isn't
Journal of Computer and System Sciences
16(3):301–322, 1978

[Jordan 60] Jordan, C.
Calculus of Finite Differences
Chelsea Publishing Company, 1960

[Knuth I] Knuth, D. E.
The Art of Computer Programming, Volume 1
Addison-Wesley, second edition 1973

[Knuth II] Knuth, D. E.
The Art of Computer Programming, Volume 2
Addison-Wesley, second edition 1981

[Knuth III] Knuth, D. E.
The Art of Computer Programming, Volume 3
Addison-Wesley, 1973

[Knuth 71] Knuth, D. E.
Mathematical Analysis of Algorithms
Proceedings of IFIP Congress 1971, 1:19–27
North-Holland, 1972

[Knuth 76a] Knuth, D. and Trabb Pardo, L.
Analysis of a Simple Factorization Algorithm
Theoretical Computer Science 3(3):321–348, 1976

[Knuth 76b] Knuth, D. E.
Big Omicron and Big Omega and Big Theta
SIGACT News 8(2):18–24, April–June 1976

[Knuth 76c] Knuth, D. E.
Mariages Stables
et leurs relation avec d'autres problèmes combinatoires
Montréal: Les Presses de l'Université de Montréal, 1976

[Knuth 78] Knuth, D. and Schönhage, A.
The Expected Linearity of a Simple Equivalence Algorithm
Theoretical Computer Science 6(3):281–315, 1978

[Knuth 89] Knuth, D. and Wilf, H.
A Short Proof of Darboux's Lemma
Applied Mathematics Letters 2(2):139–140, 1989

[Lueker 80] Lueker, G. S.
Some Techniques for Solving Recurrences
Computing Surveys 12(4):419–436, 1980

[Mertens 1874] Mertens, F.
Ein Beitrag zur analytischen Zahlentheorie
Journal für die reine und angewandte Mathematik
78:46–62, 1874

[Mil-Thom 33] Milne-Thomson, L. M.
The Calculus of Finite Differences
Macmillan, 1933

[Odlyzko 88] Odlyzko, A. and Wilf, H.
The Editor's Corner: n Coins in a Fountain
The American Mathematical Monthly 95(9):840–843, 1988

[Olver 74] Olver, F. W. J.
Asymptotics and Special Functions
Academic Press, 1974

[Page 79] Page, E. and Wilson, L.
An Introduction to Computational Combinatorics
Cambridge University Press, 1979

[Riordan 68] Riordan, J.
Combinatorial Identities
John Wiley and Sons, 1968

[Rota 75] Rota, G.
with Doubilet, Greene, Kahaner, Odlyzko, and Stanley
Finite Operator Calculus
Academic Press, 1975

[Sedgewick 75] Sedgewick, R.
Quicksort
Ph.D. Dissertation, Stanford, 1975
Garland Publishing, 1980

[Spiegel 71] Spiegel, M.
Calculus of Finite Differences and Difference Equations
Schaum's Outline Series, McGraw-Hill, 1971

[Stolarsky 77] Stolarsky, K. B.
Power and Exponential Sums of Digital Sums
Related to Binomial Coefficient Parity
SIAM Journal on Applied Mathematics 32(4):717–730, 1977

[Whittaker 40] Whittaker, E. T. and Watson, G. N.
A Course of Modern Analysis
Cambridge, 1940

[Zave 76] Zave, D. A.
A Series Expansion Involving the Harmonic Numbers
Information Processing Letters 5(1):75–77, 1976

Appendices

Appendix A: Schedule of Lectures, 1980

1 & 2 Analysis of an *in situ* permutation algorithm.
Ref: [Knuth 71]

3 Permutations with k inversions.
Generating skewed distributions.
(D. Greene, lecturer)
Ref: [Knuth III; 5.1.1–14 and 5.1.1–18]

4 & 5 Analysis of insertion sort and Shell's sort.
Ref: [Knuth III; 5.2.1]

6 The principle of postponed information (late binding).
Dijkstra's algorithm for shortest paths.
Quicksort.
Ref: [Knuth 76c], [Knuth III; 5.2.2]

7 & 8 Quicksort.
Ref: [Knuth III; 5.2.2], [Sedgewick 75]

9 & 10 Paterson's technique for hashing analysis.
Ref: Chapter 3 Operator Methods

11 Ordered hash tables.
Ref: [Amble 74]

12 Recurrence relations with minimization.

Ref: Section 2.2.1 Relations with Max or Min Functions

13 Introduction to asymptotics.

Ref: Section 4.1 Basic Concepts

14 The use of Stieltjes integration in asymptotics.

Ref: Section 4.2 Stieltjes Integration

15 Mellin transforms and the Gamma function technique.
(L. Ramshaw, lecturer)

Ref: [Knuth III, 129–134] and
work in progress by L. Guibas, L. Ramshaw, and R. Sedgewick

16 Stieltjes integration applied to a sum of reciprocal primes.

Ref: Section 4.2.3 An Example from Number Theory

17 Introduction to residue calculus.
Darboux's approach to generating functions with singularities.

Ref: Section 4.3 Asymptotics from Generating Functions

18 Saddle points and Laplace's method for obtaining asymptotics.
(D. Greene, lecturer)

Ref: Section 4.3.3 The Saddle Point Method

19 & 20 The Hungarian method.

Ref: [Erdős 59], [Erdős 60], [Knuth 76c]

Appendix B: Homework Assignments

The homework problems and their solutions appear in [Knuth III].

1	5.1.1–8 [M 24] 5.1.1–15 [M 23] 5.1.1–16 [M 25]

Show that "permutations obtainable with a stack" (namely $a_1 a_2 \ldots a_n$ of $\{1, 2, \ldots, n\}$ where $i < j < k \Rightarrow \neg(a_j < a_k < a_i)$, see exercise 2.2.1–5) can be characterized in terms of inversion tables. Find and prove a simple property of the inversion table $C_1 C_2 \ldots C_n$ that holds if and only if the permutation is obtainable with a stack. (Note: This was intended to be exercise 5.1.1–21, and in fact [Knuth III] contains the answer but not the exercise!)

2	5.2.1–5 [M 27] 5.2.1–14 [M 24] 5.2.1–37 [M 25]
3	5.2.2–7 [M 28] 5.2.2–14 [M 21] 5.2.2–20 [M 20] 5.2.2–22 [M 25]
4	6.2.1–25 [M 25] 6.2.2–6 [M 26] 6.2.2–7 [M 30]
5	6.4–27 [M 27] 6.4–34 [M 22] 6.4–49 [HM 24]
6	5.1.4–31 [HM 30] 5.2.2–57 [HM 24]
7	5.1.3–10 [HM 30] 5.2.2–54 [HM 24]
8	6.3–34 [HM 40]

Appendix C: Midterm Exam I and Solutions

Midterm Exam I

Problem 1. (a) [10 points] How many permutations on $\{1, 2, \ldots, n\}$ are sorted by at most two "bubble sort" passes? Example:

given	3	1	2	9	6	4	5	8	7
first pass	1	2	3	6	4	5	8	7	9
second pass	1	2	3	4	5	6	7	8	9

(A bubble-sort pass interchanges $K_j \leftrightarrow K_{j+1}$ iff $K_j > K_{j+1}$ for j running from 1 up to $n - 1$.)

(b) [40 points] How many permutations on $\{1, 2, \ldots, n\}$ are sorted by one double-pass of the "cocktail-shaker sort"? Example:

given	2	7	3	1	4	6	9	8	5
left-to-right pass	2	3	1	4	6	7	8	5	9
right-to-left-pass	1	2	3	4	5	6	7	8	9

(The cocktail shaker sort alternates between bubble-sort passes and similar passes in which j goes *down* from $n - 1$ to 1.)

Problem 2. Dave Ferguson's scheme for representing binary trees [exercise 2.3.1–37] would store the binary search tree

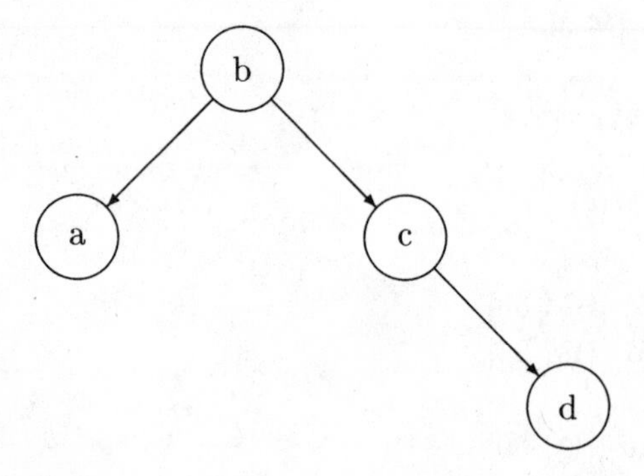

in five locations, e.g., as follows:

LOC	INFO	LINK
1	b	2
2	a	Λ
3	c	4
4		
5	d	Λ

The standard tree search and insertion procedure [Algorithm 6.2.2T] can obviously be adapted to use this representation.

Let p_{nk} be the probability that a binary search tree, constructed by starting with the empty tree and inserting n randomly ordered keys, will occupy $2n+1-2k$ locations under Ferguson's scheme; and let $P_n(z) = \sum_{k\geq 0} p_{nk} z^k$ be the corresponding generating function. For example, we have $P_1(z) = P_2(z) = z$ and $P_3(z) = \frac{2}{3}z + \frac{1}{3}z^2$.

(a) [10 points] Find a differential operator Φ_n such that $P_{n+1}(z) = \Phi_n P_n(z)$ for all $n \geq 1$.

(b) [15 points] Let D be the operator d/dz, and let U be the operator that sets $z = 1$, so that $UDP_n(z) = P_n'(1)$ is the mean value of k. Show that this mean value can be expressed as a simple function of n.

(c) [25 points] Extending the result of (b), find the variance of k as a function of n.

Problem 3. [100 points]

Consider an electric typewriter that has exactly 40 keys and an infinite carriage. The keys are:

`a b c` ... `0 1 2` ... `9` ⟨period⟩ ⟨space⟩ ⟨backspace⟩ ⟨carriage-return⟩

A monkey types at random, starting at the beginning of a line, until hitting ⟨carriage-return⟩ for the first time; this scares him, so he runs off to eat a banana.

(a) Determine the generating function $G(z) = \sum_{n\geq 0} g_n z^n$, where g_n is the number of keystroke sequences of length n that leave the word "ape" at the beginning of the line (and no other nonblank symbols).

For example, here is one such sequence of length 12 (⟨bs⟩ stands for backspace):

⟨space⟩ p ⟨bs⟩ ⟨bs⟩ ⟨bs⟩ a ⟨space⟩ e ⟨bs⟩ ⟨bs⟩ p ⟨carriage-return⟩

(Note that ⟨backspace⟩ at the beginning of a line has no effect, and characters may be overstruck.)

⇒ You need not display $G(z)$ explicitly; it suffices to specify equations that determine this function uniquely.

(b) What is the probability that the monkey types "ape" in this way? (If possible, give your answer as an explicit real number, and explain how you got it.)

[In case you dislike working on this problem, you might contemplate the probability that the monkey types certain FOUR-letter words.]

Solutions to Midterm Exam I

Solution to Problem 1.

(a) According to [Knuth III; page 108], we want to count how many inversion tables $b_1 \ldots b_n$ have all $b_j \le 2$; this is clearly $3^{n-2} \cdot 2$, for $n \ge 2$.

(b) Call the inversion table $b_1 \ldots b_n$ *easy* if a cocktail-style double-pass will sort the corresponding permutation. It turns out that there is a fairly nice way to characterize such inversion tables: $b_1 \ldots b_n$ is easy if and only if it is a valid inversion table such that either

$$\begin{aligned} & b_1 = 0 \text{ and } b_2 \ldots b_n \text{ is easy,} \\ \text{or } & b_1 = 1 \text{ and } b_2 \ldots b_n \text{ is easy,} \\ \text{or } & b_1 = 2 \text{ and } b_2 \le 1 \text{ and } b_3 \ldots b_n \text{ is easy,} \\ \text{or } & b_1 = 3 \text{ and } b_2 \le 1 \text{ and } b_3 \le 1 \text{ and } b_4 \ldots b_n \text{ is easy,} \\ \text{or } & \ldots \\ \text{or } & b_1 = n-1 \text{ and } b_2 \le 1 \text{ and } \ldots \text{ and } b_{n-1} \le 1 \text{ and } b_n \text{ is easy.} \end{aligned}$$

[*Outline of proof:* Suppose $b_1 = k > 0$. After one left-to-right pass, there are $k-1$ inversions of element 1, and at this stage the permutation must begin with $2 \ldots k\,1$ if it is to be sorted in one right-to-left pass.]

We now find that the number of easy permutations for $n \ge 2$ satisfies

$$x_n = x_{n-1} + x_{n-1} + 2x_{n-2} + 4x_{n-3} + \cdots$$

where we set $x_1 = 1$ and $x_j = 0$ for $j \le 0$. It follows that $\frac{1}{2}x_{n+1} - x_n = x_n - x_{n-1}$, i.e., $x_{n+1} = 4x_n - 2x_{n-1}$. The solution to this linear recurrence is $x_n = \frac{1}{2}\big((2+\sqrt{2}\,)^{n-1} + (2-\sqrt{2}\,)^{n-1}\big)$.

Another solution appears in [Knuth III; exercises 5.4.8–8, 9].

Solution to Problem 2.

If there are k childless nodes in the tree, Ferguson's scheme requires $2n+1-2k$ locations: one for the root and $2(n-k)$ for the children of nodes. Going from P_n to P_{n+1}, the value of k is unchanged if the new node replaces one of the $2k$ children of the k childless nodes, otherwise it increases by 1; hence

$$P_{n+1}(z) = \sum_k p_{nk}\left(\frac{2k}{n+1}z^k + \left((1-\frac{2k}{n+1}\right)z^{k+1}\right).$$

The corresponding differential operator is $\Phi_n = z + \frac{2}{n+1}z(1-z)D$; we have

$$P_n(z) = \Phi_{n-1}\ldots\Phi_0 P_0(z) \qquad \text{where } P_0(z) = 1.$$

To get the mean x_n, we note that

$$\begin{aligned} D\Phi_n &= 1 + zD + \tfrac{2}{n+1}\big((1-2z)D + z(1-z)D^2\big),\\ UD\Phi_n &= U + UD + \tfrac{2}{n+1}(-UD) = U + \tfrac{n-1}{n+1}UD. \end{aligned}$$

Hence $x_{n+1} = UDP_{n+1}(z) = (U + \frac{n-1}{n+1}UD)P_n(z) = 1 + \frac{n-1}{n+1}x_n$, and this recurrence has the solution $x_n = (n+1)/3$ for $n \ge 2$.

Similarly, to get the variance we find

$$UD^2\Phi_n = 2\tfrac{n-1}{n+1}D + \tfrac{n-3}{n+1}UD^2.$$

Let $y_n = P_n''(1)$, so that

$$y_{n+1} = 2\tfrac{n-1}{n+1}x_n + \tfrac{n-3}{n+1}y_n = \tfrac{2}{3}(n-1) + \tfrac{n-3}{n+1}y_n.$$

Applying a summation factor as on page 14, we get $z_{n+1} = (n+1)^{\underline{4}}y_{n+1} = \frac{2}{3}(n+1)^{\underline{4}}(n-1) + n^{\underline{4}}y_n = \frac{2}{3}(n+1)^{\underline{5}} + \frac{4}{3}(n+1)^{\underline{4}} + z_n$ and $z_3 = 0$. Therefore $z_n = \frac{2}{18}(n+1)^{\underline{6}} + \frac{4}{15}(n+1)^{\underline{5}}$ and $y_n = \frac{2}{18}(n+1)(n-4) + \frac{4}{15}(n+1)$ for $n \ge 4$. The variance is $y_n + x_n - x_n^2 = (n+1)\big(\frac{2}{18}(n-44) + \frac{4}{15} + \frac{1}{3} - \frac{1}{9}(n+1)\big) = \frac{2}{45}(n+1)$ for $n \ge 4$.

[*Note:* A completely different approach could also be used to get the mean and variance, using what might be called 'induction at the other end.' By considering the various choices of root nodes, we have the recurrence

$$P_n(z) = \frac{1}{n}\big(P_0(z)P_{n-1}(z) + P_1(z)P_{n-2}(z) + \cdots + P_{n-1}(z)P_0(z)\big)$$

for $n \geq 2$. Let $\boldsymbol{P}(w) = \sum_{n\geq 0} w^n P_n(z)$; this recurrence leads to the differential equation $\boldsymbol{P}' = \boldsymbol{P}^2 + P_1(z) - P_0(z)^2 = \boldsymbol{P}^2 + z - 1$, and the solution is

$$\boldsymbol{P}(w) = \sqrt{z-1}\,\tan\left(w\sqrt{z-1} + \arctan\frac{1}{\sqrt{z-1}}\right) = \frac{1+(z-1)\boldsymbol{T}(w)}{1-\boldsymbol{T}(w)},$$

where $\boldsymbol{T}(w) = \bigl(\tan w\sqrt{z-1}\bigr)/\sqrt{z-1}$. By rewriting the solution so as to avoid the square roots, we obtain

$$\boldsymbol{P}(w) = \frac{1+\sum_{k\geq 0} t_{2k+1}(z-1)^{k+1}w^{2k+1}}{1-\sum_{k\geq 0} t_{2k+1}(z-1)^k w^{2k+1}}$$

$$\text{where } \tan x = \textstyle\sum_{k\geq 0} t_{2k+1}x^{2k+1}.$$

This can be expanded in powers of $z-1$, using the values $t_1 = 1$, $t_3 = \frac{1}{3}$, $t_5 = \frac{2}{15}$, to get

$$\begin{aligned}\boldsymbol{P}(w) = \frac{1}{1-w} &+ \left(\frac{1}{3(1-w)^2} + \frac{w-1}{3}\right)(z-1)\\ &+ \left(\frac{1}{9(1-w)^3} - \frac{1}{5(1-w)^2} + \frac{1}{9} - \frac{(1-w)^3}{45}\right)(z-1)^2 + \cdots.\end{aligned}$$

So $\sum P_n(1)w^n = 1/(1-w)$, $\sum P_n'(1)w^n = \frac{1}{3}(1-w)^{-2} + \frac{1}{3}(w-1)$, $\sum \frac{1}{2}P_n''(1)w^n$ is the coefficient of $(z-1)^2$, and we find the variance in a few more steps. But this method of solution does not follow the operator approach that was specified in the problem statement.]

Solution to Problem 3.

It is convenient to consider the related function $G(x_1, x_2, x_3)$ that allows exactly x_j characters other than ⟨backspace⟩ and ⟨carriage return⟩ to be typed in column j. Then by inclusion and exclusion,

$$\begin{aligned}G = G(2,2,2) - G(2,2,1) - G(2,1,2) - G(1,2,2)&\\ + G(2,1,1) + G(1,2,1) + G(1,1,2) - G(1,1,1)&\end{aligned}$$

enumerates sequences that include all three of the letters a, p, e.

In order to avoid infinitely many equations, we consider first the set of all sequences of spaces and backspaces that begin in some column $j > 3$ and end in column $j-1$ without going left of column j until the very last step. The context-free grammar

$$L \leftarrow \langle\text{backspace}\rangle \mid \langle\text{space}\rangle\, L\, L$$

unambiguously describes such sequences, hence

$$L(z) = z + zL(z)^2$$

is the generating function $\{ z^{|\sigma|} \mid \sigma \text{ in } L \}$, and we have

$$L(z) = \bigl(1 - \sqrt{1 - 4z^2}\,\bigr)/2z = z + z^3 + 2z^5 + 5z^7 + \cdots .$$

Similarly let $Q(z)$ enumerate sequences of spaces and backspaces that begin in some column $j > 3$ and never go left of column j; the unambiguous grammar

$$Q \leftarrow \langle\text{empty}\rangle \mid \langle\text{space}\rangle\, Q \mid \langle\text{space}\rangle\, L\, Q$$

proves that

$$Q(z) = 1 + zQ(z) + zL(z)Q(z),$$
$$Q(z) = 1/\bigl(1 - z - zL(z)\bigr) = 1 + z + 2z^2 + 3z^3 + 6z^4 + \cdots .$$

[Incidentally, simple algebraic manipulations yield the identity

$$Q(z) = \bigl(1 - L(z)\bigr)/(1 - 2z),$$

a formula equivalent to $Q(z) + L(z) = 1 + 2zQ(z)$. A direct proof of the latter equation follows from the observation that every Q or L is either empty or has the form $Q\langle\text{space}\rangle$ or $Q\langle\text{backspace}\rangle$.]

Now let $G_j(z)$ be the generating function we seek when the typewriter starts j positions from the left, so that $G(z) = G_0(z)$. We have

$$G_0(z) = z + zG_0(z) + x_1 zG_1(z),$$
$$G_1(z) = z + zG_0(z) + x_2 zG_2(z),$$
$$G_2(z) = z + zG_1(z) + x_3 zG_3(z),$$

by considering sequences that begin with ⟨carriage return⟩, ⟨backspace⟩, or something else, respectively. Furthermore

$$G_3(z) = L(z)G_2(z) + Q(z)z,$$

since each sequence starting in column 4 either returns to column 3 or doesn't. The solution to this tridiagonal system of linear equations is the desired generating function $G(x_1, x_2, x_3)$.

The probability of any given sequence of keystrokes of length n is $1/40^n$, if we stop the sequence at the first ⟨carriage return⟩, and such sequences are mutually exclusive. So the probability of typing ape is $G(1/40)$.

We have now derived all that was needed to satisfy the stated problem requirements, but it is interesting to go further and coax MACSYMA to obtain reasonably simple formulas for the answer. See the attached transcript; it turns out that

$$G(z) = \frac{x_1x_2x_3z^4Q(z) - (x_1z+1)x_3z^2L(z) + (x_1-1)x_2z^3 + x_1z^2 + z}{(x_1z^3+z^2-z)x_3L(z) + x_2z^3 - (x_1+x_2)z^2 - z + 1}.$$

And after inclusion and exclusion have removed the x_i, the generating function for ape sequences begins as follows:

$$z^4 + 3z^5 + 15z^6 + 44z^7 + 163z^8 + 472z^9 + 1550z^{10} + \cdots .$$

The exact probability turns out to be

$$\frac{29996098590613938726728512750999046464990402 21\sqrt{399} - 598775887135302904116299402375695562876678654 16}{9335508254946452098018766311536840389504054535 4710},$$

which is approximately .00000042387937066 20676.

Jorge Stolfi pointed out that we could allow "o" to be typed in the second column on many typewriters, since the ink in "o" might be a subset of the ink in "p". In this case the answer would be

$$\begin{aligned} G &= G(2,3,2) - G(2,3,1) - G(2,2,2) - G(1,3,2) \\ &\qquad + G(2,2,1) + G(1,3,1) + G(1,2,2) - G(1,2,1) \\ &= z^4 + 3z^5 + 17z^6 + 52z^7 + 215z^8 + 664z^9 + 2406z^{10} + \cdots \end{aligned}$$

and the ape probability would rise to about .0000004244.

```
:macsyma

This is MACSYMA 292

FIX292 14 DSK MACSYM being loaded
Loading done

(C1) solve(L=z+z*L**2,L);

SOLVE FASL DSK MACSYM being loaded
Loading done
Solution:

                                               2
                                   SQRT(1 - 4 Z ) - 1
(E1)                           L = - ------------------
                                          2 Z

                                               2
                                   SQRT(1 - 4 Z ) + 1
(E2)                            L = ------------------
                                          2 Z
(D2)                                   [E1, E2]

(C3) solve(Q=1+z*Q+z*L*Q,Q);
Solution:

                                          1
(E3)                             Q = - -------------
                                       (L + 1) Z - 1
(D3)                                     [E3]

(C4) algebraic:true;
(D4)                                     TRUE

(C5) g0=z+z*g0+x1*z*g1;
(D5)                           G0 = G1 X1 Z + G0 Z + Z

(C6) g1=z+z*g0+x2*z*g2;
(D6)                           G1 = G2 X2 Z + G0 Z + Z

(C7) g2=z+z*g1+x3*z*g3;
(D7)                           G2 = G3 X3 Z + G1 Z + Z

(C8) g3=L*g2+z*Q;
(D8)                               G3 = Q Z + G2 L

(C9) solve([d5,d6,d7,d8],[g0,g1,g2,g3]);
```

```
Solution:

                   4                            3       2
          Q X2 Z  + ((- Q - L) X1 - Q X2) Z  - Q Z  + (Q + L) Z
(E9)  G3 = -------------------------------------------------------------
                     3                       2
         (L X1 X3 + X2) Z  + (L X3 - X2 - X1) Z  + (- L X3 - 1) Z + 1

                         4                    3                    2
              Q X2 X3 Z  + X2 (1 - Q X3) Z  + (L X3 - X2) Z  - Z
(E10) G1 = - -------------------------------------------------------------
                     3                       2
         (L X1 X3 + X2) Z  + (L X3 - X2 - X1) Z  + (- L X3 - 1) Z + 1

                             4                   3          2
                  Q X1 X3 Z  + (Q X3 + X1) Z  - Q X3 Z  - Z
(E11) G2 = - -------------------------------------------------------------
                     3                       2
         (L X1 X3 + X2) Z  + (L X3 - X2 - X1) Z  + (- L X3 - 1) Z + 1

                       4                               3                   2
          Q X1 X2 X3 Z  + (X1 (X2 - L X3) - X2) Z  + (X1 - L X3) Z  + Z
(E12) G0 = -------------------------------------------------------------
                     3                       2
         (L X1 X3 + X2) Z  + (L X3 - X2 - X1) Z  + (- L X3 - 1) Z + 1
(D12)                       [[E9, E10, E11, E12]]

(C13) g(x1,x2,x3):=([t],t:ratsimp(ev(g0,e12,e3,eval)),ratsimp(ev(t,e1)));

(D13) G(X1, X2, X3) := ([T], T : RATSIMP(EV(G0, E12, E3, EVAL)),

                                                     RATSIMP(EV(T, E1)))

(C14) g(1,1,1);

                                          Z
(D14)                                  - -------
                                         2 Z - 1

(C15) answer:g(2,2,2)-g(2,2,1)-g(2,1,2)-g(1,2,2)
+g(2,1,1)+g(1,2,1)+g(1,1,2)-g(1,1,1);

(D15)

     7                 2      6      4         6       5       4      3    2
  8 Z  + SQRT(1 - 4 Z ) (8 Z  - 4 Z ) - 12 Z  - 4 Z  - 4 Z  + 6 Z  + Z  - Z
- ---------------------------------------------------------------------------
                 7      6       5       4       3      2
             40 Z  - 4 Z  - 32 Z  - 12 Z  + 26 Z  - 3 Z  - 4 Z + 1
```

```
       7                 2         6       5         4         6         5       4       3
 + (18 Z  + SQRT(1 - 4 Z ) (2 Z  + 2 Z  - 2 Z ) + 5 Z  - 19 Z  - 6 Z  + 8 Z

    2           7         6         5        4         3       2
 + Z  - Z)/(34 Z  + 11 Z  - 44 Z  - 9 Z  + 26 Z  - 3 Z  - 4 Z + 1)

     7                 2       6       4         6       5       4       3       2
   8 Z  + SQRT(1 - 4 Z ) (4 Z  - 2 Z ) - 8 Z  + 4 Z  - 9 Z  + 4 Z  + 2 Z  - Z
 + ---------------------------------------------------------------------------
                    7       6       5        4        3     2
               16 Z  - 8 Z  - 2 Z  - 19 Z  + 20 Z  - Z  - 4 Z + 1

       7                 2     6     5     4       6       5        4       3     2
 - (6 Z  + SQRT(1 - 4 Z ) (Z  + Z  - Z ) + 5 Z  - 4 Z  - 10 Z  + 5 Z  + 2 Z

           7       6        5        4        3     2
 - Z)/(10 Z  + 7 Z  - 14 Z  - 16 Z  + 20 Z  - Z  - 4 Z + 1)

     5                 2     4     3       4       3       2
   4 Z  + SQRT(1 - 4 Z ) (Z  - Z ) - 8 Z  + 3 Z  + 2 Z  - Z
 - ------------------------------------------------------
                 5        4        3     2
             10 Z  - 21 Z  + 14 Z  + Z  - 4 Z + 1

     3               2       4       3     2        2               2       3
   2 Z  SQRT(1 - 4 Z ) + 4 Z  - 4 Z  - Z  + Z   Z  SQRT(1 - 4 Z ) - 4 Z  + Z
 - ---------------------------------------- + -----------------------------
           5        4        3     2                      3      2
       16 Z  - 24 Z  + 14 Z  + Z  - 4 Z + 1           10 Z  - 5 Z  - 2 Z + 1

     Z
 + -------
   2 Z - 1
```

```
(C16) taylor(answer,z,0,10);

HAYAT FASL DSK MACSYM being loaded
Loading done
           4       5        6        7         8         9          10
(D16)/T/ Z  + 3 Z  + 15 Z  + 44 Z  + 163 Z  + 472 Z  + 1550 Z   + . . .

(C17) ratsimp(ev(answer,z=1/40));

(D17) (299960985906139387267285127509990464649904022l SQRT(399)

 - 598775887135302904116299402375695562876678654l6)

/93355082549464520980187663115368403895040545354710

(C18) factor(denom(%));
```

```
(D18) 2 3 5 11 17 19 23 29 53 59 79 167 211 457 6673 7019 9199 20773 28559

                                                                 1291357141

(C19) bfloat(d17);

FLOAT FASL DSK MACSYM being loaded
Loading done
(D19)                          4.238793706620676B-7

(C20) time(d14,d15,d17);

TIME or [TOTALTIME, GCTIME] in msecs.:
(D20)                 [[1813, 914], [13204, 5191], [1595, 537]]
```

Acknowledgment: The MACSYMA system, developed by the Mathlab group at M.I.T., had the support of U.S. Energy Research and Development contract number E(11–1)–3070 and National Aeronautics and Space Administration grant number NSG 1323.

Appendix D: Final Exam I and Solutions

Final Exam I

Problem 1. [50 points] Find the asymptotic value of $\prod_{0\le k\le n}\binom{n}{k}$ to within a relative error of $O(1/n)$ as $n \to \infty$. [In other words, your answer should have the form $f(n)(1 + O(1/n))$ for some "explicit" function f.]

Problem 2. [100 points] Let us say that the positive integer n is *unusual* if its largest prime factor is at least $\sqrt{n}$. Thus, a prime number is unusual, as is the product of two primes. (The number 1 is also highly unusual, since it is a positive integer for which the definition makes no sense.)

Determine the asymptotic number of unusual integers n in the range $1 < n \le N$, as $N \to \infty$, with an absolute error of $O(N/(\log N)^2)$.

Hint: Count separately the unusual integers in the stated range whose largest prime factor is $\le \sqrt{N}$ [this part of the problem is worth 35 points] and those having a prime factor $> \sqrt{N}$ [this part is worth 65 points].

Additional credit will be given for answers that are correct to within $O(N/(\log N)^3)$. But you are advised to do problem 3 first before trying to get extra credit on problem 2.

Problem 3. [150 points] The following algorithm for traversing a binary tree in preorder is analogous to Algorithm 2.3.1T of [Knuth I] for inorder traversal, except that fewer items are put onto the stack:

P1. [Initialize.] Set stack `A` empty, and set the link variable $\texttt{P} \leftarrow \texttt{T}$.

P2. [$\texttt{P} = \Lambda$?] If $\texttt{P} = \Lambda$, go to step P5.

P3. [Visit `P`.] (Now `P` points to a nonempty binary tree that is to be traversed.) "Visit" `NODE(P)`.

P4. [Stack $\Leftarrow$ `RLINK(P)`.] If $\texttt{RLINK(P)} \ne \Lambda$, set $\texttt{A} \Leftarrow \texttt{RLINK(P)}$, i.e., push the value of `RLINK(P)` onto stack `A`. Then set $\texttt{P} \leftarrow \texttt{LLINK(P)}$ and return to step P2.

P5. [$\texttt{P} \Leftarrow$ Stack.] If stack `A` is empty, the algorithm terminates; otherwise set $\texttt{P} \Leftarrow \texttt{A}$ and return to step P3. ∎.

Your problem is to solve the analog of exercise 2.3.1–11 for this algorithm: What is the average value of the largest stack size occurring during the execution of Algorithm P, as a function of n, when all n-node binary trees are equally probable? Give your answer correct to within $O(n^{-1/2}\log n)$.

Solutions to Final Exam I

Solution to Problem 1.

Well, we have $\ln \prod_{0\le k\le n} \binom{n}{k} = \sum_{0\le k\le n} (\ln n! - \ln k! - \ln(n-k)!) = 2\sum_{1\le k\le n} k \ln k - (n+1)\sum_{1\le k\le n} \ln k$. By Euler's summation formula (cf. [Knuth I; exercise 1.2.11.2–7 and Eq. 1.2.11.2–18]) this is

$$\begin{aligned}
&2\left(\tfrac{1}{2}n^2 \ln n + \tfrac{1}{2}n\ln n + \tfrac{1}{12}\ln n - \tfrac{1}{4}n^2 + \ln A + O\left(\frac{1}{n^2}\right)\right)\\
&\qquad - (n+1)\left(n\ln n + \tfrac{1}{2}\ln n - n + \ln\sqrt{2\pi} + \frac{1}{12n} + O\left(\frac{1}{n^3}\right)\right)\\
&= \tfrac{1}{2}n^2 - \tfrac{1}{2}n\ln n + (1-\ln\sqrt{2\pi}\,)\,n - \tfrac{1}{3}\ln n - \tfrac{1}{12} + 2\ln A - \ln\sqrt{2\pi} + O(n^{-1}),
\end{aligned}$$

where A is Glaisher's constant. Since $\ln A = \frac{1}{12}(\gamma - \zeta'(2)/\zeta(2) + \ln 2\pi)$, a formula that can be found either in [deBruijn 70; §3.7] or by using the "Abel-Plana" formula as described in [Olver 74; §8.3.3], the answer is

$$\begin{aligned}
\exp\bigl(\tfrac{1}{2}n^2 - \tfrac{1}{2}n\ln n + (1-\tfrac{1}{2}\ln 2\pi)n - \tfrac{1}{3}\ln n - \tfrac{1}{12}&\\
+ \tfrac{1}{6}\gamma - \zeta'(2)/\pi^2 - \tfrac{1}{3}\ln 2\pi\bigr)&\bigl(1+O(1/n)\bigr).
\end{aligned}$$

Solution to Problem 2.

(a) The unusual numbers $n \le N$ whose largest prime factor is a given prime $p \le \sqrt{N}$ are the p numbers p, $2p$, …, p^2. So there are $\sum_{p\le\sqrt{N}} p$ unusual numbers of type (a). This is

$$\int_{\sqrt{2}}^{\sqrt{N}} t\,d\pi(t) = \int_{\sqrt{2}}^{\sqrt{N}} t\,dL(t) + \int_{\sqrt{2}}^{\sqrt{N}} t\,dO\bigl(t/(\log t)^{1000}\bigr)$$

where $L(t) = \int_2^t du/\ln u$. The second integral is

$$\begin{aligned}
O\bigl(\sqrt{N}\cdot\sqrt{N}/(\log\sqrt{N})^{1000}\bigr) + O\Bigl(\int_{\sqrt{2}}^{N^{1/3}} (t/(\log t)^{1000})\,dt\Bigr)+&\\
O\Bigl(\int_{N^{1/3}}^{N^{1/2}} (t/(\log t)^{1000})\,dt\Bigr)&
\end{aligned}$$

so it is $O\bigl(N/(\log N)^{1000}\bigr)$. The first integral is

$$\int_{\sqrt{2}}^{\sqrt{N}} t\,dt/\ln t = \int_2^N du/\ln u = L(N)$$

[so, curiously, $\sum_{p\le\sqrt{N}} p \approx \sum_{p\le N} 1$], which integrates by parts into the well known asymptotic form

$$N/\ln N + N/(\ln N)^2 + 2!\,N/(\ln N)^3 + \cdots + 998!\,N/(\ln N)^{999} + O\bigl(N/(\log N)^{1000}\bigr).$$

(b) The unusual numbers $n \le N$ whose largest prime factor is a given prime $p > \sqrt{N}$ are the $\lfloor N/p\rfloor$ numbers p, $2p$, $\ldots$, $\lfloor N/p\rfloor p$. So there are $\sum_{p>\sqrt{N}} \lfloor N/p\rfloor$ unusual numbers of type (b). This equals $\int_{\sqrt{N}}^{\infty} \lfloor N/t\rfloor\, d\pi(t) = \int_{\sqrt{N}}^{\infty} \lfloor N/t\rfloor\, dL(t) + \int_{\sqrt{N}}^{\infty} \lfloor N/t\rfloor\, dO\bigl(t/(\log t)^{1000}\bigr)$, and the second integral is

$$O\Bigl(N/\bigl(\log\sqrt{N}\,\bigr)^{1000}\Bigr) + O\Bigl(N\int_{\sqrt{N}}^{\infty} \frac{dt}{t(\log t)^{1000}}\Bigr)$$

so it is $O\bigl(N/(\log N)^{999}\bigr)$. The first integral is

$$\begin{aligned}\int_{\sqrt{N}}^{N} \left\lfloor \frac{N}{t} \right\rfloor \frac{dt}{\ln t} &= \int_{\sqrt{N}}^{N} \left(\frac{N}{t} - \left\{\frac{N}{t}\right\}\right)\frac{dt}{\ln t}\\ &= N\ln\ln t\,\Big|_{\sqrt{N}}^{N} - N\int_1^{\sqrt{N}} \frac{\{u\}\,du}{u^2\ln(N/u)},\end{aligned}$$

where $\{x\} = x - \lfloor x\rfloor$ and $u = N/t$. Since $\ln\ln N - \ln\ln\sqrt{N} = \ln 2$, we get

$$N\ln 2 - \frac{N}{\ln N}\int_1^{\sqrt{N}} \frac{\{u\}\,du}{u^2}\left(1 + \frac{\ln u}{\ln N} + O\left(\left(\frac{\ln u}{\ln N}\right)^2\right)\right).$$

Now $\int_1^{\infty}\bigl(\frac{\ln u}{u}\bigr)^2 du$ exists, so the O term can be dropped. We have

$$\begin{aligned}\int_1^{\sqrt{N}} \frac{\{u\}\,du}{u^2} &= \sum_{1\le k<\sqrt{N}} \int_k^{k+1} \frac{(u-k)\,du}{u^2} + O\left(\frac{1}{N}\right)\\ &= \sum_{1\le k<\sqrt{N}} \left(\ln(k+1) - \ln k - \frac{1}{k+1}\right) + O\left(\frac{1}{N}\right)\\ &= \ln\sqrt{N} - H_{\sqrt{N}} + 1 + O(1/\sqrt{N}\,)\\ &= 1 - \gamma + O(N^{-1/2}).\end{aligned}$$

Similarly (here comes the 'extra credit' part that nobody got)

$$\int_1^{\sqrt{N}} \frac{\{u\} \ln u \, du}{u^2}$$

$$= \frac{1}{2} \ln^2 \sqrt{N} + \ln \sqrt{N} - H_{\sqrt{N}} + 1 - \sum_{1 \le k < \sqrt{N}} \frac{\ln(k+1)}{k+1} + O\left(\frac{\log N}{\sqrt{N}}\right).$$

Let

$$\zeta(1+\epsilon) = \frac{1}{\epsilon} + \gamma_0 - \gamma_1 \epsilon + \frac{\gamma_2 \epsilon^2}{2!} - \frac{\gamma_3 \epsilon^3}{3!} + \cdots .$$

According to the formula in the answer to [Knuth III; exercise 6.1–8],

$$\sum_{1 \le k \le m} \frac{1}{k^{1+\epsilon}} = \zeta(1+\epsilon) + \frac{m^{-\epsilon}}{-\epsilon} + O(m^{-1-\epsilon}).$$

Expanding both sides in powers of ϵ and equating like coefficients yields

$$\sum_{1 \le k \le m} \frac{(\ln k)^r}{k} = \frac{(\ln m)^{r+1}}{r+1} + \gamma_r + O\left(\frac{(\ln m)^r}{m}\right).$$

Thus, $\int_1^{\sqrt{N}} \{u\} \ln u \, du/u^2 = 1 - \gamma_1 - \gamma_0 + O(N^{-1/2} \log N)$, and there are $N \ln 2 + (\gamma - 1)N/\ln N + (\gamma + \gamma_1 - 1)N/(\ln N)^2 + O\big(N/(\log N)^3\big)$ unusual integers of type (b); they aren't so unusual after all.

Solution to Problem 3.

Let the maximum stack size required by Algorithm P to traverse a binary tree be called its "hite," and let the analogous quantity for Algorithm T be the "height." Let $\bar{a}_{nk}$ be the number of binary trees with n nodes whose hite is at most k. If $\bar{g}_k(z) = \sum_n \bar{a}_{nk} z^n$, we find $\bar{g}_0(z) = 1/(1-z)$ and $\bar{g}_k(z) = 1 + z\bar{g}_{k-1}(z)\bar{g}_k(z) + z\bar{g}_k(z) - z\bar{g}_{k-1}(z)$. (The first term is for an empty binary tree, the next for a binary tree with left subtree hite $< k$ and right subtree hite $\le k$, and the last two are for a binary tree with left subtree hite $= k$ and an empty right subtree.) Thus

$$\bar{g}_k(z) = \frac{1 - z\bar{g}_{k-1}(z)}{1 - z\bar{g}_{k-1}(z) - z} = \cfrac{1}{1 - \cfrac{z}{1 - z\bar{g}_{k-1}(z)}},$$

and it follows immediately that $\bar{g}_k(z) = g_{2k+1}(z)$. From this surprising relation we conclude that the number of binary trees of hite k is the same as the number of binary trees of height $2k$ or $2k+1$. [It is interesting to find a one-to-one correspondence between these two sets; see the note on the last page. We don't want to spoil things for you by giving the correspondence before you've had a chance to find it for yourself, since this makes a very nice little problem.] A binary tree of height h corresponds to a binary tree of hite $\frac{1}{2}h$ or $\frac{1}{2}h - \frac{1}{2}$, so we expect the average hite to be approximately half of the average height, minus $\frac{1}{4}$. This in fact is what happens, but the point of the problem is to prove it rigorously with analytic techniques.

Following [Knuth I; exercise 2.3.1–11] and [deBruijn 72], $\bar{a}_{nk} = a_{n(2k+1)}$ is the coefficient of u^n in

$$(1-u)(1+u)^{2n}\frac{1-u^{2k+2}}{1-u^{2k+3}},$$

and $\bar{b}_{nk} = \bar{a}_{nn} - \bar{a}_{nk}$ is the coefficient of u^{n+1} in

$$(1-u)^2(1+u)^{2n}\frac{u^{2k+3}}{1-u^{2k+3}}.$$

Thus $\bar{s}_n = \sum_{k\geq 1} k(\bar{a}_{nk} - \bar{a}_{n(k-1)}) = \sum_{k\geq 0} \bar{b}_{nk}$ is the coefficient of u^{n+1} in

$$(1-u)^2(1+u)^{2n}\sum_{k\geq 0}\frac{u^{2k+3}}{1-u^{2k+3}}.$$

Let us add $\bar{b}_{n(-1)} = a_{nn}$ for convenience; $s_n + a_{nn}$ is the coefficient of u^{n+1} in

$$(1-u)^2(1+u)^{2n}\sum_{k\ \mathrm{odd}}\frac{u^k}{1-u^k},$$

which is the sum in Eq. (23) of the cited paper but with $d(k)$ replaced by $\bar{d}(k)$, the number of *odd* divisors of k. We have

$$\sum_{k\geq 1}\bar{d}(k)/k^z = \zeta(z)\Bigl(\sum_{k\ \mathrm{odd}} 1/k^z\Bigr) = \zeta(z)\bigl(\zeta(z) - 2^{-z}\zeta(z)\bigr),$$

so $\bar{s}_n + a_{nn}$ is obtained by the method in the paper except that we have an additional factor $(1-2^{b-2z})$ in the integral (29). The residue at the double pole now becomes

$$n^{(b+1)/2}\,\Gamma\bigl(\tfrac{1}{2}(b+1)\bigr)\Bigl(\frac{1}{8}\ln n + \frac{1}{8}\psi\bigl(\frac{1}{2}(b+1)\bigr) + \frac{1}{2}\gamma + \frac{1}{4}\ln 2\Bigr)$$

and at $z = -k$ it is $1 - 2^{2k+b}$ times the value in (31). The answer we seek comes to $-1 + (n+1)\bigl((-2/n)\bar{g}_0(n) + (4/n^2)\bar{g}_2(n) + O(n^{-3/2}\log n)\bigr) = \frac{1}{2}\sqrt{\pi n} - 1 + O(n^{-1/2}\log n)$.

The promised correspondence is given by the following recursive procedure:

```
ref(node) procedure transform(ref(node) value p);
begin ref(node) q, r;
if p = null then return(null);
r ← left(p); left(p) ← transform(right(p)); right(p) ← null;
if r = null then return(p);
q ← r; while true do
  begin left(q) ← transform(left(q));
  if right(q) = null then done;
  q ← right(q);
  end;
right(q) ← p; return(r);
end.
```

It can be shown that the transformed tree has the following strong property: Let s be the height of the stack when Algorithm T puts a pointer to a given node of B onto its stack, and let s' be the height of the stack after step P4 just following the time Algorithm P visits this same node in the transformed tree B'. Then $s' = \lfloor s/2 \rfloor$. Thus the stack size during the traversal of B' in preorder is almost exactly half the stack size during the inorder traversal of B, and we have a relation between the average as well as the maximum stack sizes.

Appendix E: Midterm Exam II and Solutions

Midterm Exam II

Problem 1. [50 points]

Continuing the analysis of secondary clustering in §3.4, find a "sliding operator" for Ω_m that allows $U_2G_{mn}(x)$ to be computed. Also find an analog of (3.40) that allows $U_2H_{mn}(x)$ to be computed. Express $G''_{mn}(1)$ and $H''_{mn}(1)$ as "simple" functions of m, n, P_1, and P_2, where

$$P_k = \left(1+\frac{kq}{m}\right)\left(1+\frac{kq}{m-1}\right)\ldots\left(1+\frac{kq}{m-n+1}\right).$$

Problem 2. [150 points total, distributed non-uniformly as shown below]

A student named J. H. Quick woke up one morning with an idea for a new kind of binary search tree. He had learned about the advantages of "late binding" in his studies of computer science, and he thought: "*Why should I use the first key to decide how the rest of the tree will be partitioned? I could do better by postponing that decision and letting further keys influence what happens.*" Running to his interactive workstation, he hastily prepared a file containing a description of his new data structure, which he chose to call Late Binding Trees (LBTs); and then he ate breakfast.

Unfortunately there is not room here to describe the subsequent events in Quick's life. The story about his fateful encounters with the Chuvstvenni sisters in Gstaad, who vowed to stop at nothing until they had learned his secret, will probably never be told. Let us rather turn our attention to the specifics of LBTs, suppressing the details of how this information was learned.

There are two types of nodes: branch nodes and leaves. A *branch node* contains two keys a and b, where $a < b$, and it also contains two links l and r that point to subtrees. All keys in the l subtree are $\le a$, and all keys in the r subtree are $\ge b$. Such a node can be represented by the notation '$(a\,..\,b)$', having its subtrees drawn below. A *leaf node* contains a full record, including a key a; such a node can be represented by '$[a]$'.

LBTs are never empty; they start out with a single (leaf) node. One of the nodes in the left subtree of a branch node $(a\,..\,b)$ is the leaf node $[a]$; similarly, the right subtree of $(a\,..\,b)$ always contains $[b]$. If we want to insert a new record with key x into a given LBT, we proceed as follows, assuming that x is different from all keys already in the tree:

(1) If the LBT is $[a]$, and if $a < x$, change the LBT to $(a\,..\,x)$, with left subtree $[a]$ and right subtree $[x]$. A similar construction with a and x interchanged is used if $x < a$.

(2) If the LBT has root $(a\,..\,b)$ and if $x < a$, insert the new record into the left subtree, using the same method recursively.

(3) If the LBT has root $(a\,..\,b)$ and if $x > b$, insert the new record into the right subtree, using the same method recursively.

(4) If the LBT has root $(a\,..\,b)$ and if $a < x < b$, flip a truly random coin. If it comes up heads, change the root to $(x\,..\,b)$ and insert the new record in the left subtree; otherwise change the root to $(a\,..\,x)$ and insert the new record in the right subtree.

The idea is therefore to keep track of a range of possible splitting keys in the root of the tree, instead of deciding prematurely on a particular one.

The purpose of this problem is to learn something about the analysis of algorithms by analyzing the average total external path length of LBTs, assuming that LBTs are created by inserting records in random order. The total external path length is the sum, over all leaves, of the distance from the root to the leaf. Let n be the number of leaves. Then if $n = 1$, the total external path length is always 0; if $n = 2$, it is always 2; if $n = 3$, it is always 5; and if $n = 4$, it is either 8 or 9.

(a) [15 points] Suppose that the root of an LBT is $(k\,..\,k+1)$, after inserting n keys $x_1 \ldots x_n$ that form a random permutation of $\{1, \ldots, n\}$. (In other words, the LBT starts out containing only $[x_1]$, then x_2 is inserted, and so on; there are n leaves after x_n has been inserted.) The left subtree of the root is the LBT formed by the permutation $y_1 \ldots y_k$ of $\{1, \ldots, k\}$ consisting of the x_i that are $\le k$; the right subtree is the LBT formed by the permutation $z_1 \ldots z_{n-k}$ of $\{k+1, \ldots, n\}$ consisting of the remaining x_i.

Prove that the permutations $y_1 \ldots y_k$ are not uniformly distributed; if $y_1 \ldots y_k$ has t left-to-right maxima, it occurs with probability 2^{k-t} times the probability that the identity permutation $1 \ldots k$ occurs. Similarly, the permutations $z_1 \ldots z_{n-k}$ are not uniformly random; their distribution depends on left-to-right minima.

(b) [15 points] Let p_{nk} be the probability that the root of an LBT will be $(k\,..\,k+1)$, after inserting n keys that are in uniformly random order. Find a formula for p_{nk}.

(c) [20 points] Let us say that permutations on $\{1, \ldots, n\}$ are U-distributed if all permutations are equally likely; they are L-distributed if they occur with probability proportional to 2^{-t}, where t is the number of left-to-right maxima; they are R-distributed if they occur with probability proportional to 2^{-s}, where s is the number of left-to-right minima; and they are X-distributed if they occur with probability proportional to 2^{-s-t}.

Part (a) showed that the left and right subtrees of LBTs constructed from U-distributed permutations are respectively L- and R-distributed. Prove that if we start with L-, R-, or X-distributed permutations, the subtrees are constructed from (L, X), (X, R), or (X, X)-distributed permutations, respectively.

(d) [5 points] Let U_n, L_n, R_n, and X_n be the average total external path length of the LBTs formed by distributions U, L, R, X. Prove that, for all $n \geq 2$, we have

$$U_n = n + \sum_{1 \le k < n} p_{nk}(L_k + R_{n-k}),$$
$$L_n = n + \sum_{1 \le k < n} q_{nk}(L_k + X_{n-k}),$$
$$R_n = n + \sum_{1 \le k < n} q_{n(n-k)}(X_k + R_{n-k}),$$
$$X_n = n + \sum_{1 \le k < n} r_{nk}(X_k + X_{n-k}),$$

where q_{nk} and r_{nk} are the respective probabilities that L- and X-distributed LBTs have $(k \,.\,.\, k+1)$ at the root.

(e) [20 points] Prove that $q_{nk} = \binom{k-1/2}{k-1} / \binom{n-1/2}{n-2}$ and $r_{nk} = 1/(n-1)$, for $1 \le k < n$.

(f) [5 points] Prove that $X_n = 2nH_n - 2n$.

(g) [20 points] Prove that

$$\sum_{1 \le k < n} q_{nk} X_{n-k} = \tfrac{4}{5}(n + \tfrac{1}{2})(H_{n+1/2} - H_{5/2}).$$

[*Hint:* Show that Eq. (1.47) can be used for non-integer m.]

(h) [25 points] Solve the recurrence for L_n that appears in part (d), using the repertoire method to study recurrences of the form $x_n = a_n + \sum_{1 \le k < n} q_{nk} x_k$.

(i) [25 points] Prove that $U_n = (2n + \frac{1}{2})H_n - \frac{13}{6}n - \frac{5}{12}$.

Solutions to Midterm Exam II

Solution to Problem 1.

We have $U_1 x = U_1 + U_0$ and $U_2 x = U_2 + 2U_1$ by (3.5), hence $U_0 \Omega_m = U_0$, $U_1 \Omega_m = (1 + \frac{q}{m})U_1 + pU_0$, $U_2 \Omega_m = (1 + \frac{2q}{m})U_2 + 2(p + \frac{q}{m})U_1$. Let B_m be the operator $U_2 - 2mU_1 + (m+1)mU_0$; it turns out that $B_{m+1}\Omega_m = (1 + \frac{2q}{m})B_m$. Therefore $B_{m+1}G_{mn}(x) = P_2 B_{m-n+1} x = (m-n+1)(m-n)P_2$. Furthermore $B_{m+1}G_{mn}(x) = U_2 G_{mn}(x) - 2(m+1)\bigl(m+1-(m-n)P_1\bigr) + (m+2)(m+1)$, by (3.36), and it follows that

$$U_2 G_{mn}(x) = (m-n+1)(m-n)P_2 - 2(m+1)(m-n)P_1 + m(m+1).$$

How about H? Well,

$$U_2H_{mn} = (1 + \tfrac{1}{m})U_2H_{m-1,n-1} + (\tfrac{1}{m}U_2 + \tfrac{2}{m}U_1)G_{m-1,n-1},$$

and it turns out that

$$\begin{aligned}&\tfrac{1}{m+1}U_2(H_{mn} + \tfrac{1}{p-q}G_{mn}) - \tfrac{2p}{p-q}U_1H_{mn}\\ &\qquad = \tfrac{1}{m}U_2(H_{m-1,n-1} + \tfrac{1}{p-q}G_{m-1,n-1}) - \tfrac{2p}{p-q}U_1H_{m-1,n-1}.\end{aligned}$$

Thus the quantity $\frac{1}{m+1}U_2(H_{mn} + \frac{1}{p-q}G_{mn}) - \frac{2p}{p-q}U_1H_{mn}$ turns out to be equal to $\frac{1}{m-n+1}U_2(H_{m-n,0} + \frac{1}{p-q}G_{m-n,0}) - \frac{2p}{p-q}U_1H_{m-n,0} = -\frac{2p}{p-q}$, and we can plug into our formula for U_2G_{mn} to obtain

$$U_2H_{mn} = \tfrac{m+1}{q}\bigl(m - np - 2(m-n)P_1\bigr) + \tfrac{(m+1)(m-n)}{p-q}\left(\tfrac{p}{q} - \tfrac{m-n+1}{m+1}P_2\right).$$

Note that when $p = q = \frac{1}{2}$ the latter term becomes $0/0$, so we need a separate formula for this case. By differentiating P_2 with respect to q we find that $\frac{p}{q} - \frac{m-n+1}{m+1}P_2 = (1-2q)(H_{m+1} - H_{m-n+1} + 2) + O(1-2q)^2$ as $q \to \frac{1}{2}$, hence the value of U_2H_{mn} involves harmonic numbers when $q = \frac{1}{2}$.

Solution to Problem 2.

Let us use the shorter terms "leftmax" and "leftmin" for "left to right maximum" and "left to right minimum," respectively.

(a) In order to obtain $y_1 \ldots y_k$ and $z_1 \ldots z_{n-k}$, we need $x_1x_2 = y_1z_1$ or z_1y_1, and the remaining x's must contain $y_2 \ldots y_k$ and $z_1 \ldots z_{n-k}$ merged together in some way. When x_i is being inserted, if it is a y_j we put it in the left subtree with probability $\frac{1}{2}$ if that y_j is a leftmax, and if it is a z_j we put it in the right subtree with probability $\frac{1}{2}$ if z_j is a leftmin. Otherwise the probability is 1; and if the coin flip goes the wrong way, we don't get $(k \mathinner{.\,.} k+1)$ at the root. Thus the probability of obtaining $y_1 \ldots y_k$ is proportional to 2^{-t}.

(b) For each pair of permutations $y_1 \ldots y_k$ and $z_1 \ldots z_{n-k}$ having respectively t leftmaxes and s leftmins, and for each of the $2\binom{n-2}{k-1}$ ways to merge these together as $x_1 \ldots x_n$, the probability of sending $y_1 \ldots y_k$ to the left and $z_1 \ldots z_{n-k}$ to the right is 2^{2-t-s}. Therefore p_{nk} is $2\binom{n-2}{k-1}$ times $\sum_{y,z} 2^{2-t(y)-s(z)}$, divided by $n!$.

Now the generating function for leftmaxes is

$$\textstyle\sum_y z^{t(y)} = z(1+z)\ldots(k-1+z),$$

by considering the inversion tables, hence $\sum_y 2^{1-t(y)} = (k-\frac{1}{2})^{\underline{k-1}}$. It follows that

$$p_{nk} = 2\binom{n-2}{k-1}(k-\tfrac{1}{2})^{\underline{k-1}}(n-k-\tfrac{1}{2})^{\underline{n-k-1}}/n! = \binom{k-1/2}{k-1}\binom{n-k-1/2}{n-k-1}\Big/\binom{n}{2}.$$

Ken Clarkson also found the curious formula

$$p_{nk} = 8\binom{k-1/2}{n}\binom{n-2}{k-1}(-1)^{n-k}.$$

(c) The leftmins in $x_1 \ldots x_n$ all occur in $y_1 \ldots y_k$, and the leftmaxes all occur in $z_1 \ldots z_{n-k}$, except perhaps for the very first ones. Thus, the probability of obtaining a particular permutation y is equal to $2\binom{n-2}{k-1}$ times $\sum_z 2^{2-t(y)-s(z)}p(x)$, where $p(x)$ is the probability that $x_1 \ldots x_n$ is input. If we assume (as we may) that $x_1 < x_2$, then $p(x)$ is proportional to $2^{-t(z)}$, $2^{-s(y)}$, $2^{-t(z)-s(y)}$ in distributions L, R, X. The result is proportional to $2^{-t(y)}$, $2^{-s(y)-t(y)}$, $2^{-s(y)-t(y)}$, so the left subtrees have distributions L, X, X. The right subtrees are similar.

(d) The total path length is n plus the total path length of the left subtree plus the total path length of the right subtree. So, with probability p_{nk}, we obtain a contribution of $n+L_k+R_{n-k}$ to the average total path length. The duality between left and right shows that $q_{k(n-k)}$ is the probability that an R-distributed LBT has $(k\,..\,k+1)$ at the root. It follows that $L_n = R_n$ (which was obvious).

(e) By part (c), the probability q_{nk} is proportional to the double sum $\sum_{y,z} 2^{-t(y)-s(z)-t(z)}\binom{n-2}{k-1}$, where the constants of proportionality for fixed n are independent of k. The generating function $\sum_y z^{s(y)+t(y)}$ is equal to $(z^2)(2z)(1+2z)\ldots(k-2+2z)$, hence $\sum_y 2^{-s(y)-t(y)} = \frac{1}{4}(k-1)!$. Thus q_{nk} is proportional to $\binom{k-1/2}{k-1}$ and r_{nk} is independent of k; it only remains to find the constants of proportionality so that $\sum q_{nk} = \sum r_{nk} = 1$. See equation (1) below.

(f) We have $X_n = C_{n-1}$ in standard quicksort [GKP; (2.12), (2.14)].

(g) We have $(1-z)^{-1-m} = \sum_{n\ge 0}\binom{n+m}{n}z^n$, for all complex m, by the binomial theorem. Differentiating with respect to m (this idea was suggested by John Hobby), we obtain (1.47):

$$(1-z)^{-1-m}\ln(1-z)^{-1} = \sum_{n\ge 0}\binom{n+m}{n}(H_{n+m}-H_m)z^n.$$

Let us now tabulate a bunch of formulas that follow immediately from this identity, since the formulas will prove useful in the sequel. All sums are over the range $1 \le k < n$. We use the facts that $\binom{k-1/2}{k-1}(k-1) = \frac{3}{2}\binom{k-1/2}{k-2}$, that $\binom{k-1/2}{k-1}$ is the coefficient of z^{k-1} in $(1-z)^{-3/2}$, etc.

$$\sum \binom{k-1/2}{k-1} = \binom{n-1/2}{n-2} \tag{1}$$

$$\sum \binom{k-1/2}{k-1}(k-1) = \frac{3}{2}\binom{n-1/2}{n-3} \tag{2}$$

$$\sum \binom{k-1/2}{k-1}(H_{k-1/2} - H_{1/2}) = \binom{n-1/2}{n-2}(H_{n-1/2} - H_{3/2}) \tag{3}$$

$$\sum \binom{k-1/2}{k-1}(n-k)(H_{n-k} - H_1) = \binom{n+1/2}{n-2}(H_{n+1/2} - H_{5/2}) \tag{4}$$

$$\sum \binom{k-1/2}{k-1}(k-1)(H_{k-1/2} - H_{3/2}) = \frac{3}{2}\binom{n-1/2}{n-3}(H_{n-1/2} - H_{5/2}) \tag{5}$$

$$\sum \binom{k-1/2}{k-1}\binom{n-k-1/2}{n-k-1}(k-1) = \frac{3}{2}\binom{n}{n-3} \tag{6}$$

$$\sum \binom{k-1/2}{k-1}\binom{n-k-1/2}{n-k-1}(H_{k-1/2} - H_{1/2}) = \binom{n}{n-2}(H_n - H_2) \tag{7}$$

$$\sum \binom{k-1/2}{k-1}\binom{n-k-1/2}{n-k-1}(k-1)(H_{k-1/2} - H_{3/2}) = \frac{3}{2}\binom{n}{n-3}(H_n - H_3) \tag{8}$$

Each of these identities is obtained by looking at the coefficients of the product of two generating functions.

The answer to part (g) comes from (4), after multiplying by $2/\binom{n-1/2}{n-2}$.

(h) We need to solve $L_n = n + \frac{4}{5}(n+\frac{1}{2})(H_{n+1/2} - H_{5/2}) + \sum q_{nk} L_k$, for $n \ge 2$. Trying $x_n = n-1$ in $x_n = a_n + \sum q_{nk} x_k$ gives $a_n = \frac{2}{5}n + \frac{1}{5}$ for $n \ge 2$, by (2), since $n - 1 = a_n + \frac{3}{2}\binom{n-1/2}{n-3}/\binom{n-1/2}{n-2} = a_n + \frac{3}{5}(n-2)$. Similarly, trying $x_n = H_{n-1/2} - H_{1/2}$ gives $a_n = \frac{2}{3}$, by (3); and $x_n = (n-1)(H_{n-1/2} - H_{3/2})$ gives $a_n = \frac{2}{5}(n+\frac{1}{2})(H_{n-1/2} - H_{5/2}) + \frac{2}{5}(n-1)$ by (5). Taking an appropriate linear combination of all this yields the solution $L_n = (2n + \frac{1}{4})(H_{n-1/2} - H_{1/2}) - \frac{5}{6}(n-1)$.

(i) We have $U_n = n + 2\sum p_{nk} L_k$. Write $L_k = 2(k-1)(H_{k-1/2} - H_{3/2}) + \frac{9}{4}(H_{k-1/2} - H_{1/2}) + \frac{1}{2}(k-1)$ and use (8), (7), (6), to get

$$U_n = n + 2(n-2)(H_n - \tfrac{11}{6}) + \tfrac{9}{2}(H_n - \tfrac{3}{2}) + \tfrac{1}{2}(n-2).$$

We may conclude that LBTs do not deserve to be implemented; they offer us instructive insights into discrete mathematics and the analysis of algorithms, but they will never become known as Quicksearch. It is somewhat surprising that $U_n \le L_n \le X_n$, since a reluctance to insert "extreme" elements might be thought to make the inequalities go the other way.

Appendix F: Final Exam II and Solutions

Final Exam II

Problem 1. [75 points]

Find the asymptotic value of $S_n = \sum_{0<k<n} H_k/(2n-k)$, correct to terms of $O(n^{-3/2})$.

Problem 2. [100 points total]

Let a_n be the number of paths from $(0,0)$ to (n,n) on a grid, where we are allowed to go at each step from (i,j) to $(i,j+1)$ or $(i+1,j)$ or $(i+1,j+1)$. Thus, $(a_0, a_1, a_2, a_3, \ldots) = (1, 3, 13, 63, \ldots)$.

(a) [50 points] Let $A(z) = \sum_n a_n z^n$. Use the method of (4.125) to prove that $A(z) = 1/\sqrt{1-6z+z^2}$.

(b) [50 points] Find the asymptotic value of a_n as $n \to \infty$, giving explicit values of constants c, p, and θ such that $a_n = cn^p\theta^n + O(n^{p-1}\theta^n)$.

Problem 3. [125 points total]

A certain professor gives final exams that contain an infinite number of problems. In order to solve each problem, the student must have solved all of the preceding problems perfectly, since the professor stops grading an exam as soon as he finds a mistake. Each student has probability p of getting any particular problem right, independently of the other students, and independently of the problem number. For example, if $p = \frac{1}{2}$, there is probability 2^{-n-1} that a particular exam will have exactly n problems right.

The professor gives an A$^+$ to the student who solves the most problems, provided that only one student had the maximum score. Otherwise nobody in the class gets A$^+$.

(a) [25 points] Write down an expression for the probability that an A$^+$ is given when n students take the exam. (Your expression can be left in the form of a summation, since there appears to be no "closed form" for the probability in question.)

(b) [100 points] Find the asymptotic behavior of the probability that an A$^+$ is given after n students take the exam, for fixed p as $n \to \infty$. Assume that $0 < p < 1$.

Important note: You must solve problem 3a correctly if you want to get any credit for problem 3b. Make sure that your formula gives the value $2p(1+p)^{-1}$ when $n = 2$ and the value $3p(1+p^2)(1+p)^{-1}(1+p+p^2)^{-1}$ when $n = 3$, before you tackle the asymptotics.

Solutions to Final Exam II

Solution to Problem 1.

Summing by parts yields

$$2S_n = \sum_{0<k<2n} H_k/(2n-k) - H_n^2,$$

which equals $H_{2n}^2 - H_{2n}^{(2)} - H_n^2$ by (1.48). Now

$$H_{2n}^2 - H_n^2 = \left(\ln n + \ln 2 + \gamma + \tfrac{1}{4n} + O(n^{-2})\right)^2 - \left(\ln n + \gamma + \tfrac{1}{2n} + O(n^{-2})\right)^2,$$

and $H_{2n}^{(2)} = \frac{1}{6}\pi^2 - \frac{1}{2n} + O(n^{-2})$ by [Knuth III; exercise 6.1–8]. Multiplying out and collecting terms yields

$$(\ln 2)(\ln n) + \gamma \ln 2 + \frac{1}{2}(\ln 2)^2 - \frac{1}{12}\pi^2$$
$$-\frac{1}{4}n^{-1}\ln n + \frac{1}{4}n^{-1}(\ln 2 + 1 - \gamma) + O(n^{-2}\log n).$$

[This problem was too easy. It would have been better to ask for the asymptotics of, say, $\sum_{1\le k\le n-\sqrt{n}} H_k/(2n-k)$. Then the asymptotics could be worked out most easily by using the identity $\sum_{1\le k\le m} H_k/(n-k) = \sum_{1\le k\le m} H_{n-k}/k - H_m H_{n-m-1}$.]

Solution to Problem 2.

Set $F(w,z) = \sum a_{mn} w^m z^n$, where a_{mn} is the number of paths from $(0,0)$ to (m,n). Then $F = 1 + wF + zF + wzF$, so we have $F(w,z) = (1-w-z-wz)^{-1}$. The diagonal terms are

$$A(z) = \frac{1}{2\pi i}\oint F(t, z/t)\frac{dt}{t} = \frac{1}{2\pi i}\oint \frac{dt}{t - t^2 - z - zt}.$$

The denominator can be written in factored form, $-(t-r(z))(t-s(z))$, where $r(z) = \frac{1}{2}\left(1 - z + \sqrt{1-6z+z^2}\right)$ and $s(z) = \frac{1}{2}\left(1 - z - \sqrt{1-6z+z^2}\right)$.

Let $|z|$ be small, so that $r(z)$ is near 1 and $s(z)$ is near 0. Integrate around a contour with small $|t|$ that encloses the point $s(z)$; then make $|z|$ and $|s(z)|$ even smaller so that $|z/t|$ is small enough to guarantee absolute convergence of $\sum a_{mn} t^m (z/t)^n$. (It is clear that $a_{mn} \le 3^{m+n}$, so such a contour exists.) The result is $A(z) = \text{residue at } s(z) = 1/\big(r(z) - s(z)\big)$.

Now $A(z) = 1/\sqrt{(1-\theta z)(1-\phi z)}$, where $\theta = 3+\sqrt{8}$ and $\phi = 3-\sqrt{8}$. Let $w = \theta z$ and $\alpha = \phi/\theta$ so that $A(z) = B(w) = 1/\sqrt{(1-w)(1-\alpha w)} = \sum (a_n/\theta^n) w^n$. We therefore want to find the asymptotics of the coefficients of the inverse of (4.108). We have $1/\sqrt{1-\alpha w} = 1/\sqrt{1-\alpha-\alpha(w-1)} = (1-\alpha)^{-1/2} + (w-1)R(w)$ where R is analytic for $|w| < \alpha^{-1}$, so the coefficients r_n of R are $O(\beta^{-n})$ for some $\beta > 1$. Thus $B(w) = (1-\alpha)^{-1/2}(1-w)^{-1/2} + (1-w)^{1/2}R(w)$, where the latter term is $\sum \binom{1/2}{k}(-w)^k \sum r_m w^m$; it follows as in (4.114) that its nth coefficient is $O(n^{-3/2})$. The nth coefficient of the first term is $(1-\alpha)^{-1/2}\binom{-1/2}{n}(-1)^n = (1-\alpha)^{-1/2}\binom{n-1/2}{n}$, which is of order $n^{-1/2}$, so $a_n = \theta^n (1-\alpha)^{-1/2}\binom{n-1/2}{n} + O(a_n/n)$.

We have $\binom{-1/2}{n}(-1)^n = 2^{-2n}\binom{2n}{n}$, and Stirling's approximation tells us that this is $1/\sqrt{\pi n} + O(n^{-3/2})$. Thus the desired answer is

$$a_n = \frac{1+\sqrt{2}}{2^{5/4}\sqrt{\pi n}}(3+\sqrt{8})^n + O\big((3+\sqrt{8})^n n^{-3/2}\big).$$

Incidentally, the numbers a_{mn} arise in surprisingly many contexts. We have, for example, $a_{mn} = \sum \binom{m}{k}\binom{n+k}{m} = \sum \frac{(m+n-k)!}{(m-k)!(n-k)!k!} = \sum \binom{m}{k}\binom{n}{k}2^k = \sum \binom{k}{m}\binom{k}{n}2^{-1-k}$. Also, a_{mn} is the number of different n-tuples of integers $(x_1,\ldots,x_n)$ such that $|x_1| + \cdots + |x_n| \le m$; this is the volume of a sphere of radius m in the n-dimensional "Lee metric."

Solution to Problem 3.

The probability that a particular student gets A$^+$ with exactly m problems correct is the probability of scoring m (namely $p^m(1-p)$) times the probability that each other student missed at least one of the first m problems (namely $(1-p^m)^{n-1}$). Multiplying by n, since each student has the chance for an A$^+$, we obtain $A_n^+ = n(1-p)\sum_{m\ge 0} p^m(1-p^m)^{n-1}$. (Similar formulas arise in the analysis of radix exchange sorting in [Knuth III, 5.2.2], when $p = \frac{1}{2}$, and in the more general treatment of exercise 6.3–19.)

Let $Q_n = nA_{n+1}^+(n+1)^{-1}(1-p)^{-1} = \sum_{m\ge 0} np^m(1-p^m)^n$. Let $x = np^m$; the summand is $x(1-x/n)^n$, which is $xe^{-x}\big(1+O(x^2/n)\big)$ when $x \le n^\epsilon$. Let $T_n = \sum_{m\ge 0} np^m e^{-np^m}$. We have $Q_n - T_n = X_n + Y_n$ where

$$\begin{aligned} X_n &= \textstyle\sum_{m\ge 0, np^m \ge n^\epsilon} np^m\big((1-p^m)^n - e^{-np^m}\big) \\ &= \textstyle\sum_{m\ge 0, np^m \ge n^\epsilon} np^m O(e^{-np^m}) = O(n\log n\, e^{-n^\epsilon}) \end{aligned}$$

is exponentially small, since $1-p^m \le e^{-p^m}$ and there are $O(\log n)$ terms. Also

$$Y_n = \textstyle\sum_{m\ge 0, np^m < n^\epsilon} np^m\bigl((1-p^m)^n - e^{-np^m}\bigr)$$
$$= \textstyle\sum_{m\ge 0, np^m < n^\epsilon} np^m O\bigl(e^{-np^m}(np^m)^2/n\bigr),$$

which is $O(n^{3\epsilon-1})$ since it reduces to a geometric series after we use the obvious upper bound $e^{-np^m} \le 1$. Applying the Gamma function method, we have

$$T_n = \sum_{m\ge 0} \frac{1}{2\pi i}\int_{1/2-i\infty}^{1/2+i\infty} np^m\Gamma(z)(np^m)^{-z}\,dz$$
$$= \frac{n}{2\pi i}\int_{1/2-i\infty}^{1/2+i\infty} \frac{\Gamma(z)n^{-z}}{1-p^{1-z}}dz,$$

(cf. Eq. 5.2.2–45), which can be evaluated as the negative of the sum of the integrand's residues at its poles in the right half plane. Thus

$$T_n = -\frac{1}{\ln p} - \frac{2}{\ln p}\sum_{k\ge 1}\Re\bigl(\Gamma(1+2\pi ik/\ln p)\exp(-2\pi ik\ln n/\ln p)\bigr) + O(n^{-M})$$

for arbitrary M. The quantity in the sum is bounded since it is periodic in n (note that it has the same value at n and pn). So we can say that $A_n^+ = (1-p)/\ln(\frac{1}{p}) + f(n) + O(\frac{1}{n})$, where $f(n)$ is a certain periodic function. The absolute value of $f(n)$ is extremely small unless p is extremely small, since $\Gamma(1+ti) = O(t^{1/2}e^{-\pi t/2})$; and each term of $f(n)$ has average value zero, so $f(n)$ is zero on the average. But $f(n)$ is present and it is not $o(1)$. One might suspect that A_n^+ would approach 0 or 1 when $n\to\infty$, so the result is a bit surprising.

Exercise 5.2.2–54 gives another approach to the answer, by which we obtain the convergent series

$$A_n^+ = \frac{1-p}{\ln(1/p)}\Bigl(1 + 2n\sum_{k\ge 1}\Re\bigl(B(n, 1+\tfrac{2\pi ik}{\ln p})\bigr)\Bigr).$$

The Beta function in this sum has the asymptotic value

$$n^{-1-ibk}\Gamma(1+ibk)\bigl(1-\tfrac{1}{2}(ibk+b^2k^2)n^{-1}+O(n^{-2})\bigr),$$

where $b = 2\pi/\ln p$; so we obtain the periodic function mentioned above, as well as the coefficient of n^{-1}. (It appears that exercise 5.2.2–54 should be mentioned much more prominently in the next edition of [Knuth III].)

Appendix G: Midterm Exam III and Solutions

Midterm Exam III

Problem 1. Let C_n be the nth Catalan number,

$$C_n = \binom{2n}{n}\frac{1}{n+1},$$

and let $Q_n = C_n(H_{2n} - H_{n+1})$. Thus we have the following values for small n:

$n =$	0	1	2	3	4	5
$C_n =$	1	1	2	5	14	42
$Q_n =$	-1	0	$\frac{1}{2}$	$\frac{11}{6}$	$\frac{73}{12}$	$\frac{1207}{60}$

Prove the amazing identity

$$\sum_{k=0}^{n} C_k Q_{n-k} = Q_{n+1} - \binom{2n+1}{n}\frac{1}{n+1}.$$

Hint: Consider the derivative of

$$B(z)^x = \sum_{n\geq 0}\binom{2n+x}{n}\frac{x}{2n+x}z^n, \qquad B(z) = \frac{1-\sqrt{1-4z}}{2z}$$

with respect to x.

Problem 2. Given $0 \leq m < F_{n+2}$, the *Fibonacci representation* of m is defined to be $(d_n \ldots d_2 d_1)_F$ if $m = d_n F_{n+1} + \cdots + d_2 F_3 + d_1 F_2$, where each d_k is 0 or 1 and $d_k d_{k+1} = 0$ for $1 \leq k < n$. The *Fibonacci permutation of* order n is the permutation of $\{0, 1, \ldots, F_{n+2} - 1\}$ that is obtained by reflecting the representations of $0, 1, \ldots, F_{n+2} - 1$ from right to left. For example, here are the representations and their reflections when $n = 4$:

m	representation	reflection	permuted value	inversions
0	$(0000)_F$	$(0000)_F$	0	0
1	$(0001)_F$	$(1000)_F$	5	0
2	$(0010)_F$	$(0100)_F$	3	1
3	$(0100)_F$	$(0010)_F$	2	2
4	$(0101)_F$	$(1010)_F$	7	0
5	$(1000)_F$	$(0001)_F$	1	4
6	$(1001)_F$	$(1001)_F$	6	1
7	$(1010)_F$	$(0101)_F$	4	3

Let X_n be the total number of inversions in the Fibonacci permutation of order n. (When $1 \le n \le 5$ we have $X_n = 0, 1, 3, 11, 32$, respectively.) Find a closed form for X_n in terms of Fibonacci numbers. (Do not use the number $(1+\sqrt{5})/2$ explicitly in your answer.)

Problem 3. The Eikooc Monster is a dual to Paterson's Cookie Monster: The probability that it *doesn't* grow is proportional to its size. More precisely, if E has eaten k cookies before a new cookie is thrown, it eats the new cookie with probability $1 - pk$. (Monster C absorbs cookies that fall on it, while E eats those that it can see in the rest of the yard.) The differential operator Θ corresponding to E is

$$x + p(1-x)xD.$$

Find a family of eigenoperators for Θ corresponding to the book's family $V_1, V_2, \ldots$. Use your operators to deduce the mean and variance of the number of distinct coupons collected after n purchases of random coupons drawn uniformly and independently from the set $\{1, 2, \ldots, m\}$.

Derive asymptotic formulas for the mean and variance of this number when $n = cm$, for fixed c, correct to $O(1)$ as m and $n \to \infty$.

Problem 4. Find the mean and variance of the number of comparisons when the following sorting algorithm is applied to n distinct inputs $a[1\,..\,n]$:

```
procedure pokeysort (n: integer);
    begin if n > 1 then
        repeat Set k to random element of {1, 2, ..., n};
        Exchange a[k] ↔ a[n];
        pokeysort (n − 1);
        until a[n − 1] ≤ a[n];
    end.
```

Solutions to Midterm Exam III

Solution to Problem 1.

[The class presented a variety of interesting approaches; here is yet another, which includes several formulas that may be handy in other investigations.] If we write $C_{-1} = -1$, we have $B(z) = \sum_{n\ge0} C_n z^n$ and $B(z)^{-1} = -\sum_{n\ge0} C_{n-1} z^n$. The derivative of the hinted formula is

$$B(z)^x \ln B(z) = \sum_{n\ge0} \binom{2n+x}{n} \frac{z^n}{2n+x} \left(1 + x(H_{2n+x-1} - H_{n+x})\right).$$

The special case $x = 0$ gives $\ln B(z) = \sum_{n\ge1} \binom{2n}{n} \frac{1}{2n} z^n$; the special case $x = 1$ gives $B(z) \ln B(z) = \sum_{n\ge1}(C_n + Q_n)z^n$. Multiplying by $B(z)^y$ and equating coefficients of z^n gives

$$\sum_{k=0}^{n} \binom{2k+x}{k} \frac{1}{2k+x}\bigl(1 + x(H_{2k+x-1} - H_{k+x})\bigr)\binom{2n-2k+y}{n-k} \frac{y}{2n-2k+y}$$
$$= \binom{2n+x+y}{n} \frac{1}{2n+x+y}\bigl(1 + (x+y)(H_{2n+x+y-1} - H_{n+x+y})\bigr).$$

Set $x = 1$ and $y = -1$, getting $\sum_{k=0}^{n}(C_k + Q_k)(-C_{n-1-k}) = \binom{2n}{n} \frac{1}{2n}$. But when $n > 0$ we have $\sum_{k=0}^{n} C_k C_{n-1-k} = 0$, hence $\sum_{k=0}^{n+1} Q_k C_{n-k} = -\binom{2n+2}{n+1} \frac{1}{2n+2} = -\binom{2n+1}{n} \frac{1}{n+1}$.

Solution to Problem 2.

[This was in part an exercise in mastering complexity without getting mired in details.] Everybody solved this problem by deriving a recurrence, usually with the idea that $X_n = X_{n-1} + X_{n-2} + C_n$ where C_n is the number of "cross inversions" between the first block of F_{n+1} values and the last block of F_n values. The value of C_n can be written in several ways, notably $C_n = Y_n + Z_{n-1} + Y_{n-2} + Z_{n-3} + \cdots = C_n + Y_n + Z_{n-1} + C_{n-2}$, where $Y_n = \binom{F_n}{2}$, $Z_n = Y_n + F_n$, and $C_0 = C_1 = 0$. It turns out that $C_n = \frac{1}{2} F_{n-1}(F_{n+2} - 1)$. Jim Hwang made the interesting observation that the inversion table entries $B_0B_1B_2\ldots$ begin the same for each n; therefore it would be of interest to study the partial sums $B_0 + B_1 + \cdots + B_{m-1}$ as a function of m.

But there's another interesting approach to the solution, based directly on the binary representations: Each inversion corresponds to strings α, β, β', γ of 0s and 1s such that

$$(\alpha\, 0\, \beta\, 1\, \gamma)_F < (\alpha\, 1\, \beta'\, 0\, \gamma)_F\,, \qquad (\alpha\, 0\, \beta\, 1\, \gamma)_F^R > (\alpha\, 1\, \beta'\, 0\, \gamma)_F^R\,.$$

(If $i < j$ and $a_i > a_j$, the Fibonacci forms of i and j must differ first in 0 versus 1, last in 1 versus 0.) The number of such pairs with $|\alpha| = k$, $|\beta| = n - k - l$, and $|\gamma| = l$, is $F_{k+1}F_{n-k-l-1}^2 F_{l+1}$; hence X_n is the sum of this quantity over $0 \le k, l \le n$.

Let $F(z) = \sum F_{k+1}z^k = 1/(1 - z - z^2)$ and

$$G(z) = \sum_{k\ge0} F_{k+1}^2 z^k = \frac{1}{5}\left(\frac{3 - 2z}{1 - 3z + z^2} + \frac{2}{1 + z}\right).$$

Then X_n is $[z^{n-2}]\,F(z)^2G(z)$. The partial fraction expansion of this generating function involves the denominators $(1-\phi z)^2$, $(1-\phi z)$, $(1-\hat{\phi}z)^2$, $(1-\hat{\phi}z)$, $(1-\phi^2 z)$, $(1-\hat{\phi}^2 z)$, and $1+z$. Hence there must be seven constants such that

$$X_n = (\alpha n+\beta)F_n + (\gamma n+\delta)F_{n+1} + \epsilon\, F_{2n} + \zeta F_{2n+1} + \eta(-1)^n\,.$$

MACSYMA quickly determines these constants when given the values of X_n for $1 \le n \le 7$, using '`solve`'. [Incidentally, '`solve`' is much quicker than '`partfrac`' followed by additional manipulation.] The answer is

$$X_n = \frac{7F_{2n+1} + 4F_{2n} - (4n+15)F_{n+1} + (2n+7)F_n + 8(-1)^n}{20}\,.$$

Incidentally, a random permutation of this size would have exactly $I_n = \frac{1}{4}\,F_{n+2}(F_{n+2}-1) = \frac{1}{20}\bigl(7F_{2n+1}+4F_{2n}-2(-1)^n-5F_{n+1}-5F_n\bigr)$ inversions on the average. The Fibonacci permutation is "pretty random" by the inversion-count criterion, since $X_n - I_n$ is of order $\sqrt{I_n}\log I_n$.

Solution to Problem 3.

We have $U_n x = nU_{n-1} + U_n$, hence $U_n(1-x) = -n\,U_{n-1}$. Let's search for an eigenoperator of the form $U_n x^{n-a}$: We have

$$\begin{aligned}
U_n x^{n-a}\Theta &= U_n x^{n-a}\bigl(x + p(1-x)xD\bigr)\\
&= U_n x^{n+1-a} + p\,U_n(1-x)x^{n+1-a}D\\
&= U_n x^{n+1-a} - pn\,U_{n-1}\bigl(Dx^{n+1-a} - (n+1-a)x^{n-a}\bigr)\\
&= (1-pn)U_n x^{n+1-a} + pn(n+1-a)U_{n-1}x^{n-a}\\
&= (1-pn)(U_n x^{n-a} + n\,U_{n-1}x^{n-a}) + pn(n+1-a)U_{n-1}x^{n-a}\\
&= (1-pn)U_n x^{n-a} + (n+pn-pna)U_{n-1}x^{n-a}\,.
\end{aligned}$$

Therefore we get an eigenoperator with eigenvalue $1-pn$ when $a = 1+p^{-1}$.

The formula $U_n f(x) = \sum_k \binom{n}{k} f^{(k)}(1)U_{n-k}$ tells us that the eigenoperator $U_n x^{n-a}$ can be written $\sum_k \binom{n}{k}(n-1-p^{-1})^{\underline{k}}U_{n-k}$. It is convenient to normalize it so that the coefficient of U_0 is $+1$; then $V_nG(z) = 1$ when $G(z) = 1$. With this normalization (suggested by Arif Merchant), we have

$$V_n = \sum_k \binom{n}{k}\frac{(-1)^k}{(p^{-1})^{\underline{k}}}U_k$$

and therefore

$$U_n = (p^{-1})^{\underline{n}}\sum_k \binom{n}{k}(-1)^k V_k\,.$$

If $G_n(z) = \Theta^n(1)$, the mean and variance are now easily found to be respectively $p^{-1}\bigl(1-(1-p)^n\bigr)$ and $p^{-2}(1-p)(1-2p)^n + p^{-1}(1-p)^n - p^{-2}(1-p)^{2n}$, in agreement with the answer to [Knuth III; exercise 5.2.5–5] when $p = 1/m$.

When $n = cm$, the mean is $(1 - e^{-c})m + O(1)$; the variance reduces to $(e^c - 1 - c)e^{-2c}m + O(1)$, fairly small.

Solution to Problem 4.

The probability generating function $G_n(z)$ is defined by the recurrence $G_1(z) = 1$, $G_n(z) = z\,G_{n-1}(z)\bigl(\frac{1}{n} + \frac{n-1}{n}\,G_n(z)\bigr)$ for $n > 1$. Hence

$$G_n(z) = F_n\bigl(z\,G_{n-1}(z)\bigr), \qquad F_n(z) = \frac{z}{n - (n-1)z}.$$

Now [Knuth I; exercise 1.2.10–17] tells us that

$$\begin{aligned}
\operatorname{Mean}(G_n) &= \operatorname{Mean}(F_n)\bigl(1 + \operatorname{Mean}(G_{n-1})\bigr) = n\bigl(1 + \operatorname{Mean}(G_{n-1})\bigr)\\
\operatorname{Var}(G_n) &= \operatorname{Var}(F_n)\bigl(1 + \operatorname{Mean}(G_{n-1})\bigr)^2 + \operatorname{Mean}(F_n)\operatorname{Var}(G_{n-1})\\
&= \frac{n-1}{n}\operatorname{Mean}(G_n)^2 + n\operatorname{Var}(G_{n-1}).
\end{aligned}$$

Dividing these recurrences by $n!$ leads to sums such as

$$\operatorname{Mean}(G_n) = \sum_{1\le k<n} \frac{n!}{k!} = S_n - n! - 1$$

where S_n has a convenient closed form:

$$S_n = \sum_{0\le k\le n} \frac{n!}{k!} = n!\,e - e\gamma(n+1, 1) = \lfloor n!\,e \rfloor.$$

The variance can also be expressed in terms of S_n:

$$\operatorname{Var}(G_n) = (S_n - n! - n - 1)^2 + 3S_n - n! - n^2 - 2n - 3.$$

Incidentally, we can "solve" the recurrence for $G_n(z)$ and write

$$G_n(z) = z^n / H_n(a), \qquad H_n(z) = n!\left(z + \sum_{k=2}^{n}\left(\frac{1}{k!} - \frac{1}{(k-1)!}\right) z^k\right);$$

then $\operatorname{Var}(G_n) = -\operatorname{Var}(H_n)$, and the latter can be calculated directly.

Appendix H: Final Exam III and Solutions

Final Exam III

Problem 1. A certain gambler starts with \$1 at time 0. If he has \$$k$ at time t, he makes k independent fair bets of \$1 each, double or nothing; this determines his capital at time $t+1$. (Thus, at time 1, he is equally likely to be broke or to have \$2. At time 2, he has (\$0, \$2, \$4) with probability $(\frac{5}{8}, \frac{2}{8}, \frac{1}{8})$.) (a) Find the mean and variance of his capital at time t. (b) Find the asymptotic probability p_t that he is broke at time t, with absolute error $O(t^{-2})$. *Hint:* Consider the quantity $1/(1-p_{t+1}) - 1/(1-p_t) - 1/2$.

Problem 2. Each of n parallel processors is executing a random process such that, if the processor is running at time t, it is stopped at time $t+1$ with probability p (independent of all other circumstances). Once stopped, a processor remains stopped. Find the asymptotic value of the expected waiting time until all processors have stopped, assuming that they are all running at time 0. Your answer should be correct to $O(n^{-1})$.

Problem 3. Find the asymptotic value of $S_n = \sum_{1\le k<n/2} e^{-k^2/n} k^{-2}$, with absolute error $O(n^{-10})$. (This is the quantity $r_{-2}(n/2)$ in [Knuth III; Eq. 5.2.2–35].)

Problem 4. Last year's AA qual included a problem in which it was shown that the number of ways to arrange n coins into a so-called "fountain" has the generating function

$$\sum_{n\ge 0} f_n z^n = \cfrac{1}{1-\cfrac{z}{1-\cfrac{z^2}{1-\cfrac{z^3}{\cdots}}}} = \frac{P(z)}{Q(z)}$$

where

$$P(z) = \sum_{k\ge 0} \frac{(-1)^k z^{k(k+1)}}{(1-z)(1-z^2)\ldots(1-z^k)},$$

$$Q(z) = \sum_{k\ge 0} \frac{(-1)^k z^{k^2}}{(1-z)(1-z^2)\ldots(1-z^k)}.$$

Prove that $Q(z)$ has exactly one root ρ in the disk $|z| \le .6$, and that $f_n = c\rho^{-n} + O(.6^{-n})$ for some nonzero constant c. Use MACSYMA to evaluate

ρ^{-1} and c to several decimal places. *Hints:* First express the quantity $(1-z)(1-z^2)(1-z^3)Q(z)$ in the form $A(z)+R(z)$, where $A(z)$ is a polynomial of degree 9 and $|R(z)|$ is small when $|z|<1$. Find the roots of $A(z)$ using the `allroots` command. Then find a radius r such that $A(z)$ has exactly one root for $|z|\le r$ and such that $|R(z)|<|A(z)|$ for $|z|=r$. Then apply Rouché's theorem.

Solutions to Final Exam III

Solution to Problem 1.

(a) $G_0(z)=z$; $G_{t+1}(z)=f\bigl(G_t(z)\bigr)$, where $f(z)=\frac{1}{2}+\frac{1}{2}z^2$. Since f has mean and variance 1, [Knuth I; exercise 1.2.10–17] implies that $\mathrm{Mean}(G_t)=1$ and $\mathrm{Var}(G_t)=t$.

(b) Let $\epsilon_t=(1-p_t)/2$. We have $p_t=G_t(0)$, so $p_{t+1}=\frac{1}{2}+\frac{1}{2}p_t^2$. Hence $\epsilon_{t+1}=\epsilon_t(1-\epsilon_t)$, and we have

$$\frac{1}{\epsilon_{t+1}}=\frac{1}{\epsilon_t}+1+\frac{1}{\epsilon_t^{-1}-1}\,,\qquad \epsilon_0=\frac{1}{2}\,.$$

A bootstrapping argument now shows that $\epsilon_t^{-1}\ge t+2$, hence

$$\epsilon_t^{-1}\le t+2+\sum_{k=0}^{t-1}\frac{1}{k+1}=t+2+H_t\,;$$

hence

$$\begin{aligned}\epsilon_t^{-1}&\ge t+2+\sum_{k=0}^{t-1}\frac{1}{k+1+H_k}\\ &=t+2+\sum_{k=0}^{t-1}\left(\frac{1}{k+1}+O\left(\frac{\log k}{k^2}\right)\right)=t+2+H_t+O(1)\,.\end{aligned}$$

We have proved that $\epsilon_t^{-1}=t+\ln t+O(1)$; hence $p_t=1-2/\bigl(t+\ln t+O(1)\bigr)=1-2t^{-1}+2t^{-2}\ln t+O(t^{-2})$. [Let $C=\lim_{t\to\infty}(\epsilon_t^{-1}-t-H_t)$. Is it possible to go further and estimate the quantity $\delta_t=\epsilon_t^{-1}-t-H_t-C$, as $t\to\infty$?]

Solution to Problem 2.

Let $q=1-p$ and $Q=1/q$. The probability that the processors are not all stopped at time t is $R_t=1-(1-q^t)^n$, so the expected waiting time is $W=\sum_{t\ge0}R_t=1-\sum_{t\ge1}\sum_{k\ge1}\binom{n}{k}(-1)^k q^{tk}=1-\sum_{k\ge1}\binom{n}{k}(-1)^k/(Q^k-1)$.

We proceed as in [Knuth III; exercise 5.2.2–54] to represent the sum as

$$\frac{(-1)^n\, n!}{2\pi i} \oint \frac{dz}{z(z-1)\,\ldots\,(z-n)\,(Q^z-1)}$$

where the contour encircles $\{1,\ldots,n\}$ and no other poles. If we increase the contour to a large rectangle whose upper and lower segments have imaginary part $\pm 2\pi(N+\frac{1}{2})/\ln Q$ where N is an integer, the contour integral approaches zero, so the sum of the residues inside approaches zero. The residue at 0 is the coefficient of z in

$$\frac{1}{(1-z)(1-\frac{1}{2}z)\,\ldots\,(1-\frac{1}{n}z)(1+\frac{1}{2}z\ln Q+\cdots)\ln Q},$$

namely $(H_n - \frac{1}{2}\ln Q)/\ln Q$. The sum of residues at $1,\ldots,n$ is $1-W$. And the sum of residues at $\ln Q + ibm$ and $\ln Q - ibm$, where $b = 2\pi/\ln Q$ and $m \geq 1$, is $1/\ln Q$ times twice the real part of

$$\frac{n!}{(ibm)(ibm+1)\,\ldots\,(ibm+n)} = B(n+1, ibm) = \Gamma(ibm)\, n^{\underline{ibm}}$$

$$= \Gamma(ibm)\, n^{ibm}\bigl(1 + O(n^{-1})\bigr)\,.$$

(The last estimate comes by expanding $n^{\underline{ibm}}$ in terms of generalized Stirling numbers; for example, we have

$$n^{\underline{\alpha}} = n^{\alpha}\left(\begin{bmatrix}\alpha\\ \alpha\end{bmatrix} - \begin{bmatrix}\alpha\\ \alpha-1\end{bmatrix} n^{-1} + \begin{bmatrix}\alpha\\ \alpha-2\end{bmatrix} n^{-2} + O(n^{-3})\right)\,.$$

See [GKP; exercise 9.44].) Now $|\Gamma(ibm)n^{ibm}| = O(e^{-\pi m/2})$, so we have

$$W = \frac{H_n}{\ln Q} + \frac{1}{2} + \frac{2}{\ln Q}\sum_{m\geq 1} \Re\bigl(\Gamma(ibm)n^{ibm}\bigr) + O(n^{-1})\,.$$

The sum is a bounded function $f(n)$ that is periodic in the sense that $f(n) = f(Qn)$. Tomás Feder used Euler's summation formula to deduce the remarkable representation

$$f(n) = \int_0^{\infty} \left(\left(\frac{\log u/n}{\log Q}\right)\right) e^{-u}\, du$$

where $((x))$ is the sawtooth function [Knuth II; §3.3.3].

Solution to Problem 3.

Let $g(x) = (e^{-x^2} - 1)/x^2$ and $f(x) = g(x/\sqrt{n}\,)$. Then

$$\begin{aligned}
nS_n &- 1 - nH^{(2)}_{n-1} \\
&= \sum_{0 \le k < n} f(k) + O(e^{-n/4}) \\
&= \int_0^n f(x)\,dx + \sum_{j=1}^{18} \frac{B_j}{j!} f^{(j-1)}(x)\bigg|_0^n + O(n^{-9}) \\
&= \sqrt{n} \int_0^{\sqrt{n}} g(y)\,dy + \sum_{j=1}^{18} \frac{B_j}{j!} n^{-(j-1)/2} g^{(j-1)}(y)\bigg|_0^{\sqrt{n}} + O(n^{-9})\,.
\end{aligned}$$

Consider first

$$\begin{aligned}
\int_0^{\sqrt{n}} g(y)\,dy &= \frac{1 - e^{-y^2}}{y}\bigg|_0^{\sqrt{n}} - 2\int_0^{\sqrt{n}} e^{-y^2}\,dy \\
&= \frac{1}{\sqrt{n}} - 2\int_0^{\infty} e^{-y^2}\,dy + O(e^{-n})\,.
\end{aligned}$$

We also have

$$g^{(j-1)}(\sqrt{n}\,) = (-1)^j 2^{\overline{j-1}} n^{-(j+1)/2} - O(e^{-n})$$

and $g^{(j-1)}(0)$ is nonzero only when j is odd, so we can square it for $j > 1$. Thus

$$nS_n - 1 - nH^{(2)}_{n-1} = 1 - \sqrt{\pi n} - \frac{1}{2}\left(-\frac{1}{n} + 1\right) + \sum_{j=1}^{9} \frac{B_{2j}}{n^{2j}} + O(n^{-9})\,.$$

Also

$$H^{(2)}_{n-1} = \frac{\pi^2}{6} - \frac{1}{n} - \frac{1}{2n^2} - \sum_{j=1}^{9} \frac{B_{2j}}{n^{2j+1}} + O(n^{-10})$$

by Euler's summation formula. Finally, therefore,

$$S_n = \frac{\pi^2}{6} - \sqrt{\frac{\pi}{n}} + \frac{1}{2n} + O(n^{-10})\,.$$

The error is, in fact, $O(n^{-1000000})$. (Check: $S_{10} = 1.134434895$; the approximation yields 1.134434945.)

There are (at least) two other ways to solve this problem: We can use the Gamma-function approach, as pointed out in [Knuth III; exercise 5.2.2–51]; or we can use the Poisson summation formula (found, for example, in [Henrici II]).

Solution to Problem 4.

(The hints are due to A. Odlyzko.) We have

$$A(z) = 1 - 2z - z^2 + z^3 + 3z^4 + z^5 - 2z^6 - z^7 - z^9$$

and

$$R(z) = \frac{z^{16}}{1-z^4}\left(1 - \frac{z^9}{1-z^5} + \frac{z^9}{1-z^5}\frac{z^{11}}{1-z^6} - \cdots\right).$$

If $|z| = r < 1$ we have

$$|R(z)| \le \frac{r^{16}}{1-r^4}\left(1 + \frac{r^9}{1-r^5} + \left(\frac{r^9}{1-r^5}\right)^2 + \cdots\right) = \frac{r^{16}}{1-r^4}\frac{1-r^5}{1-r^5-r^9}.$$

The roots of $A(z) = 0$ are, in increasing order of magnitude:

$$\begin{array}{ll} r_1 \approx .58 & \\ r_2, r_3 \approx .75 \pm .08i & |r_2| = |r_3| \approx .75 \\ r_4, r_5 \approx -.47 \pm .82i & |r_4| = |r_5| \approx .94 \\ r_6, r_7 \approx -1.06 \pm .37i & |r_6| = |r_7| \approx 1.12 \\ r_8, r_9 \approx .49 \pm 1.58i & |r_8| = |r_9| \approx 1.66\,. \end{array}$$

To apply Rouché's theorem, we want to find a value of r such that $|A(z)|$ is relatively large when $|z| = r$ but $|R(z)|$ is relatively small. The hard part is to show that $|A(z)| = |z - r_1| \ldots |z - z_9|$ is relatively large.

One idea is to observe that

$$|z-a+bi|\,|z-a-bi| \ge \begin{cases} (r-|a|)^2 + b^2, & \text{if } a(r^2+a^2+b^2) > 2r(a^2+b^2); \\ \min((r-|a|)^2 + b^2, \big|b(r^2-a^2-b^2)\big|/|a+ib|), & \\ \qquad \text{otherwise.} & \end{cases}$$

(The proof is by setting $z = re^{i\theta}$ and taking the derivative with respect to θ. Extrema occur when $\sin\theta = 0$ or when we have $\cos\theta = a(r^2 + a^2 + b^2)/(2r(a^2 + b^2))$.) Unfortunately this idea isn't enough by itself; the product of all these bounds turns out to be less than r^{16}.

Better bounds are possible if we use the inequality $|z - r_k| \ge \big||r_k - r| - |z - r|\big|$. Then if $|r_k - r| > .5$ we can conclude that $|z - r_k| \ge |r_k - r| - .5$, whenever $|z - r| \le .5$; similarly if $|r_k - r| < .5$ we can conclude that $|z - r_k| \ge .5 - |r_k - r|$, whenever $|z - r| \ge .5$.

Putting these ideas together yields a rigorous proof that $|A(z)| > |R(z)|$ for all z on the circle $|z| = r$, for any choice of r between .59 and .68. (See the attached MACSYMA transcript. The computed values $r_1, \ldots, r_9$

are only approximations to the true roots of $A(z)$; but the fact that the difference $(z-r_1)\dots(z-r_9)-A(z)$ has very small coefficients implies that our calculations are plenty accurate when $|z|\le 1$.)

Consequently Rouché's theorem applies, and $Q(z)$ has exactly one root ρ_0 inside $|z|=r$. This root is real, and Newton's method converges quickly to

$$\rho_0 = 0.57614876914275660229786\ldots .$$

The contour integral

$$\frac{1}{2\pi i}\oint_{|z|=r}\frac{P(z)\,dz}{Q(z)\,z^{n+1}}$$

is $O(r^{-n})$, and the sum of residues inside is

$$f_n + \frac{P(\rho_0)}{\rho_0^{n+1}\,Q'(\rho_0)}\,.$$

Hence we have $f_n = c_0\rho_0^{-n} + O(r^{-n})$, where $c_0 = P(\rho_0)/\bigl(\rho_0 Q'(\rho_0)\bigr)$; numerically

$$\frac{1}{\rho_0} = 1.7356628245303472565826\ldots ;$$

$$c_0 = 0.3123633245967414530662279\ldots .$$

It turns out that the next largest root of $Q(z)$ is also real; it is

$$\rho_1 = .81559980\,;$$

$$c_1 = P(\rho_1)/\bigl(\rho_1 Q'(\rho_1)\bigr) = .03795269\,.$$

The graph of $Q(z)$ looks like this for $.5\le z\le .9$:

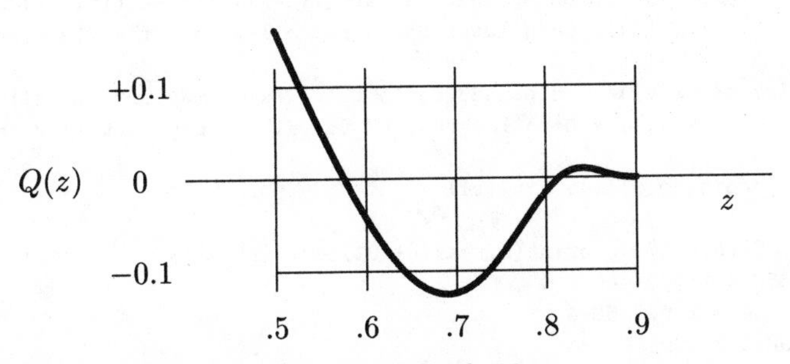

There is another root between .88 and .89.

To check, Odlyzko computed $f_{120} = 17002133686539084706594617194$, and found that $f_{120} - c_0/\rho_0^{120} \approx 1.6\times 10^9$. If we subtract c_1/ρ_1^{120} the error goes down to 1.3×10^5. (Odlyzko's work was published in [Odlyzko 88] after this exam was given.)

```
This is MACSYMA 304
(C1) t(k,z):=z↑(k↑2)/prod(1-z↑j,j,1,k);

(C2) q(n,z):=sum((-1)↑k*t(k,z),k,0,n);

(C3) a:num(factor(q(3,z)));
                 9     7       6     5       4     3     2
(D3)             Z  + Z  + 2 Z  - Z  - 3 Z  - Z  + Z  + 2 Z - 1

(C4) allroots(a);

(C5) for n thru 9 do print(n,r[n]:rhs(part(d4,n)),abs(r[n]));
1 0.575774066 0.575774066
2 0.81792161 %I - 0.469966464 0.94332615
3 - 0.81792161 %I - 0.469966464 0.94332615
4 0.07522564 %I + 0.74832744 0.75209896
5 0.74832744 - 0.07522564 %I 0.75209896
6 0.36716983 %I - 1.05926119 1.1210923
7 - 0.36716983 %I - 1.05926119 1.1210923
8 1.58184962 %I + 0.493013173 1.65689777
9 0.493013173 - 1.58184962 %I 1.65689777

(C6) rmax(r):=r↑16/(1-r↑4)/(1-r↑9/(1-r↑5));

(C7) bound1(a,b,r):=block([t,s],s:a↑2+b↑2,t:(r-abs(a))↑2+b↑2,
      if a*(r↑2+s)>2*r*s then t else min(t,abs(b*(r↑2-s))/sqrt(s)));

(C8) bound2(a,b,r):=block([t,s],s:abs(a+b*%i-r),t:bound1(a,b,r),
      if s<.5 then t else max(t,(.5-s)↑2));

(C9) bound3(a,b,r):=block([t,s],s:abs(a+b*%i-r),t:bound1(a,b,r),
      if s>.5 then t else max(t,(s-.5)↑2));

(C10) amin1(r):=(r-r[1])*prod(bound2(realpart(r[2*k]),imagpart(r[2*k]),r),
                 k,1,4) + 0*"a lower bound for all z such that |z-r|>=.5";

(C11) amin2(r):=(r-r[1])*prod(bound3(realpart(r[2*k]),imagpart(r[2*k]),r),
                 k,1,4) + 0*"a lower bound for all z such that |z-r|<=.5";

(C12) amin(r):=min(amin1(r),amin2(r));

(C13) for n:58 thru 70 do print(n,rmax(n*.01),amin(n*.01));
58 1.86410865E-4 1.40064462E-4
59 2.4762821E-4 4.7992996E-4
60 3.2769739E-4 8.2895893E-4
61 4.320998E-4 1.18362144E-3
62 5.6784198E-4 1.54014562E-3
63 7.438718E-4 1.89452055E-3
64 9.7160927E-4 2.24249464E-3
65 1.26562865E-3 2.57957187E-3
66 1.6445353E-3 2.90100428E-3
```

```
67 2.13209912E-3 3.19357002E-3
68 2.75873208E-3 3.17922997E-3
69 3.56342027E-3 3.1048643E-3
70 4.596277E-3 2.92984536E-3

(C14) qprime(n,z):=sum((-1)↑k*t(k,z)*logtprime(k,z),k,0,n);

(C15) logtprime(k,z):=k↑2/z+sum(j*z↑(j-1)/(1-z↑j),j,1,k);

(C16) loop(z):=block([zo,zn],zo:0,zn:z,
      while abs(zo-zn)>10↑-10 do(zo:zn,print(zn:iterate(zo))),zo);

(C17) t(8,.59)+0*"an upper bound on the alternating sum Q(.59)-Q(8,.59)";
(D17)                              1.3545675E-14

(C18) iterate(z):=bfloat(z-q(8,z)/qprime(8,z));

(C19) loop(5.8B-1);
5.761132798756077B-1
5.761487662923891B-1
5.761487691427566B-1
5.761487691427566B-1
(D19)                         5.761487691427566B-1

(C20) p(n,z):=sum((-1)↑k*t(k,z)*z↑k,k,0,n);

(C21) c(rho):=-p(8,rho)/(rho*qprime(8,rho));

(C22) c(d19);
(D18)                         3.123633245967415B-1

(C23) expand(prod(z-r[k],k,1,9)-d3);
                              8                       8                        7
(D23) - 1.49011612E-8 %I Z  - 7.4505806E-9 Z  - 1.49011612E-8 %I Z

                        6                   6                         5
 + 1.49011612E-8 %I Z  + 8.9406967E-8 Z  + 2.98023224E-8 %I Z

                    5                      4                     4
 + 1.63912773E-7 Z  + 1.1920929E-7 %I Z  - 1.78813934E-7 Z

                        3                  3                         2
 - 4.47034836E-8 %I Z  - 2.98023224E-7 Z  - 1.04308128E-7 %I Z

                    2
 - 2.01165676E-7 Z  - 5.2154064E-8 %I Z + 1.49011612E-7 Z + 7.4505806E-9 %I

(C24) "The sum of the absolute values of those coefficients is
  an upper bound on the difference between the true A(z) and
  the polynomial that is bounded by amin";
```

Appendix I: A Qualifying Exam Problem and Solution

Qual Problem

The result of a recent midterm problem was to analyze LBTs and to show that their average path length is about the same as that of ordinary binary search trees.

But shortly after the midterm was graded, our sources discovered that Quick was undaunted by that analysis. According to reliable reports, he has recently decided to try salvaging his idea by including new information in each node.

The nodes in Quick's new data structures, which he calls ILBTs (Improved Late Binding Trees), contain a size field that tells how many leaves are in the subtree rooted at that node. Step (4) on page 102 is now replaced by a new step: When a branch node is being split, the insertion continues in whichever subtree is currently smaller. (If the subtree sizes are equal, a random decision is made as before.)

The purpose of this problem is to carry out a "top level" analysis of Quick's new algorithm. Let p_{nk} be the probability that the root is $(k\,..\,k{+}1)$ after inserting a random permutation of $\{1,\ldots,n\}$. (We assume that all permutations of the x's are equally likely; first x_1 is made into an ILBT by itself, then x_2 through x_n are inserted one by one.) Let $P_{nk} = n!\,p_{nk}$. Then it can be verified that we have the following values of P_{nk} for $1 \le k < n$ and $1 \le n \le 6$:

$n = 2$:	2				
$n = 3$:	3	3			
$n = 4$:	6	12	6		
$n = 5$:	18	42	42	18	
$n = 6$:	72	162	252	162	72

(a) Find a recurrence relation that defines the numbers P_{nk}.

(b) Let $Q_{nk} = 2P_{nk}\max(k, n-k)/\bigl(k!(n-k)!\bigr)$, so that we have the following triangle:

$n = 2$:	4				
$n = 3$:	6	6			
$n = 4$:	6	12	6		
$n = 5$:	6	21	21	6	
$n = 6$:	6	27	42	27	6

Show that for most values of n and k the numbers Q_{nk} satisfy the same recurrence as Pascal's triangle, i.e., $Q_{nk} = Q_{(n-1)k} + Q_{(n-1)(k-1)}$. Find all the exceptions, and state the recurrence obeyed at the exceptional points.

(c) Let $a_k = Q_{(2k)k}$. Prove that for $k > 1$,

$$a_k = \sum_{1 \le j < k} \frac{2j+1}{j} a_j c_{k-j},$$

where c_n is the number of binary trees with n external nodes.

(d) Let $B(z) = \frac{1}{2}(1 + \sqrt{1-4z})$ and $C(z) = \frac{1}{2}(1 - \sqrt{1-4z})$, so that $B(z) + C(z) = 1$, $B(z) - C(z) = \sqrt{1-4z}$, $B(z)C(z) = z$, and $C(z)^2 = C(z) - z$; recall that $C(z)$ is the generating function $c_1 z + c_2 z^2 + c_3 z^3 + \cdots$ for binary trees. Let $f_k = a_k/k$, and set up the generating function $F(z) = f_1 z + f_2 z^2 + \cdots$. Convert the recurrence in part (c) to a differential equation for F, and solve this equation to obtain a "closed form" for a_k. [*Possible hint:* Show that the derivative of $B(z)F(z)$ has a simple form.]

(e) Apply the recurrence of part (b) to the generating function $Q(w,z) = \sum_{k,n} Q_{nk} w^k z^{n-k}$, and use the values of a_k found in part (d) to obtain a formula for $Q(w,z)$ as an explicit function of w and z.

(f) Find a "simple" expression for the coefficient of $w^n z^{n+r}$ in the power series for $\sqrt{1-4wz}/(1-w-z)$, when $r \ge 0$. [*Hint:* Consider the problem for fixed r and variable n. You may wish to use the identity $C(z)^s/\sqrt{1-4z} = \sum_n \binom{2n+s}{n} z^{n+s}$ and the facts about $B(z)$ and $C(z)$ that are stated in (d).]

(g) Show that, therefore,

$$p_{nk} = \frac{1}{2}\left(\frac{k+1}{n-k} - \frac{k}{n-k+1}\right) - \frac{1}{2n} + \frac{2k}{n(n-1)} \qquad \text{for } 1 \le k < \tfrac{1}{2}n.$$

Note: Do NOT simply take this formula or an equivalent one and prove it by induction. You should present a scenario that explains how you could have discovered this solution by yourself in a systematic manner without lucky guesses.

Qual Solution

(a) If $x_1 \ldots x_n$ is a permutation of $\{1, \ldots, n\}$, let $\bar{x}_1 \ldots \bar{x}_{n-1}$ be the permutation of $\{1, \ldots, n-1\}$ that arises when the elements of $x_1 \ldots x_{n-1}$ that exceed x_n are reduced by 1. The permutation $x_1 \ldots x_n$ leads to the root $(k \,..\, k+1)$ if and only if one of the following happens: (1) $x_n < k$ and $\bar{x}_1 \ldots \bar{x}_{n-1}$ leads to the root $(k-1 \,..\, k)$. (2) $x_n = k$ and $\bar{x}_1 \ldots \bar{x}_{n-1}$ leads to the root $(k-1 \,..\, k)$ and either $k-1 < n-k$ or ($k-1 = n-k$ and a random coin flip comes up heads). (3) $x_n = k+1$ and $\bar{x}_1 \ldots \bar{x}_{n-1}$ leads to the root $(k \,..\, k+1)$ and either $k > n-1-k$ or ($k = n-1-k$ and a random coin flip comes up tails). (4) $x_n > k+1$ and $\bar{x}_1 \ldots \bar{x}_{n-1}$ leads to the root $(k \,..\, k+1)$. Therefore we find, for $1 \le k < n$ and $n > 2$,

$$\begin{aligned} P_{nk} = P_{(n-1)(k-1)}\bigl(k-1+[n+1>2k]+\tfrac{1}{2}[n+1=2k]\bigr) \\ + P_{(n-1)k}\bigl(n-k-1+[n-1<2k]+\tfrac{1}{2}[n-1=2k]\bigr). \end{aligned}$$

(b) It is easy to see that $P_{nk} = P_{n(n-k)}$, so $Q_{nk} = Q_{n(n-k)}$. Thus it suffices to consider $k \le n-k$. If $k < n-k-1$, the above recurrence reads

$$\begin{aligned} \frac{Q_{nk}k!(n-k)!}{2(n-k)} = \frac{Q_{(n-1)(k-1)}(k-1)!(n-k)!}{2(n-k)}(k-1+1) \\ + \frac{Q_{(n-1)k}k!(n-1-k)!}{2(n-k-1)}(n-k-1), \end{aligned}$$

i.e., $Q_{nk} = Q_{(n-1)(k-1)} + Q_{(n-1)k}$. If $k = n-k$, it reads

$$\begin{aligned} \frac{Q_{nk}k!\,k!}{2k} = \frac{Q_{(n-1)(k-1)}(k-1)!\,k!}{2k}(k-1+1) \\ + \frac{Q_{(n-1)k}k!(k-1)!}{2k}(k-1+1), \end{aligned}$$

so Pascal's relation holds again. But if $k = n-k-1$, we have

$$\begin{aligned} \frac{Q_{nk}k!(k+1)!}{2(k+1)} = \\ \frac{Q_{(n-1)(k-1)}(k-1)!(k+1)!}{2(k+1)}(k-1+1) + \frac{Q_{(n-1)k}k!\,k!}{2k}(k+\tfrac{1}{2}), \end{aligned}$$

hence $Q_{nk} = Q_{(n-1)(k-1)} + Q_{(n-1)k} + \frac{1}{n-1}Q_{(n-1)k}$. By symmetry, if $k = n-k+1$ we have $Q_{nk} = Q_{(n-1)(k-1)} + Q_{(n-1)k} + \frac{1}{n-1}Q_{(n-1)(k-1)}$. Pascal's relation therefore holds except when $(n,k) = (2,1)$ or when $|n-2k| = 1$.

(c) It is convenient to tip the triangle sideways and to associate Q_{nk} with the point $(k, n-k)$ in a grid. We can interpret Q_{nk} as 4 times the sum, over all paths from $(1,1)$ to $(k, n-k)$, of the products of the weights of the edges, where edges run from (i,j) to $(i+1,j)$ and to $(i,j+1)$; the weight of such an edge is 1, except when $i=j$ it is $1+1/(2j)$. Now a_k is 4 times the sum over paths from $(1,1)$ to (k,k), so we can break the sum into various sub-sums depending on the greatest diagonal point (j,j) on the path, for $j<k$. The jth sub-sum is a_j times $1+1/(2j)$ times the number of subpaths from (j,j) to (k,k) that do not touch the diagonal, since all edge weights but the first are 1 on such subpaths. There are $2c_{k-j}$ such subpaths.

(d) Since $kf_k = \sum_j (2j+1) f_j c_{k-j} + 4[k=1]$, we have $zF'(z) = 4z + C(z)\bigl(2zF'(z) + F(z)\bigr)$, and this simplifies to

$$z\sqrt{1-4z}\,F'(z) = 4z + C(z)F(z).$$

Following the hint, which follows from the general method of finding an integrating factor for first-order differential equations, we find

$$\begin{aligned}\bigl(B(z)F(z)\bigr)' &= B(z)F'(z) - F(z)/\sqrt{1-4z}\\ &= \frac{B(z)}{z\sqrt{1-4z}}\bigl(z\sqrt{1-4z}\,F'(z) - C(z)F(z)\bigr)\\ &= 4B(z)/\sqrt{1-4z} = 2/\sqrt{1-4z} + 2.\end{aligned}$$

Thus $B(z)F(z) = 2C(z) + 2z$, and in a few more steps we find the solution $a_n = 2n(c_n + c_{n+1}) = n\binom{2n}{n}\bigl(\frac{1}{2n-1} + \frac{2}{n+1}\bigr)$, for $n \ge 1$.

(e) $(1-w-z)Q(z) = \sum w^k z^{n-k}(Q_{nk} - Q_{(n-1)k} - Q_{(n-1)(k-1)}) = 4wz + \frac{1}{2}(w+z)(f_1 wz + f_2 w^2 z^2 + f_3 w^3 z^3 + \cdots)$, hence we have

$$Q(w,z) = \frac{4wz + \frac{1}{2}\bigl(w^{-1} - w + z^{-1} - z - (w^{-1} + w + z^{-1} + z)\sqrt{1-4wz}\,\bigr)}{1-w-z}.$$

(f) The coefficient of $w^n z^{n+r}$ in $g(wz)h(w,z)$ is the coefficient of x^n in $g(x)h_r(x)$, if $h_r(x) = \sum a_{m(m+r)} x^m$ and $h(w,z) = \sum a_{mn} w^m z^n$, since multiplication by $g(wz)$ affects only the coefficients having the same exponent offset. Hence the coefficient of $w^n z^{n+r}$ in $\sqrt{1-4wz}/(1-w-z)$ is the coefficient of x^n in $\sqrt{1-4x}\sum\binom{2n+r}{n}x^n = \bigl(C(x)/x\bigr)^r = C(x)^r\bigl(B(x) - C(x)\bigr)x^{-r}/\sqrt{1-4x} = \bigl(C(x)/x\bigr)^{r-1}/\sqrt{1-4x} - x\bigl(C(x)/x\bigr)^{r+1}/\sqrt{1-4x} = \sum\bigl(\binom{2n+r-1}{n} - \binom{2n+r-1}{n-1}\bigr)x^n$.

(g) For $r > 0$, the coefficient of $w^n z^{n+r}$ in $Q(w,z)$ can now be computed by considering the various terms in part (e). Let $b_r = \binom{2n+r}{n}$. Then $Q_{(2n+r)n} = 4(b_{r-1} - b_{r-2}) + \frac{1}{2}\bigl(\binom{2n+r+1}{n+1} - b_r + b_{r-1} + b_{r+1} - b_{r-1} - \binom{2n+r}{n+1} + b_r - b_{r-1} + b_{r-2} + b_r - 2b_{r-1} + b_{r-2} - b_r + b_{r+1} - b_r - b_{r-2} + b_{r-1} - b_{r-2}\bigr) = b_{r+1} + 3b_{r-1} - 4b_{r-2}$. Multiply by $\frac{1}{2}(n+r-1)!\,n!/(2n+r)!$ to get $p_{(2n+r)n}$. A different formula applies when $r = 0$, because of the w^{-1} and z^{-1} terms.

Final comment: A note to J. H. Quick. "When $x = k/n < \frac{1}{2}$ we have $p_{nk} \approx \frac{1}{2}\bigl((1-x)^{-2} - 1\bigr) + 2x$, hence ILBT's do a reasonably good job of partitioning. The distribution of permutations in the left and right subtrees is not random, and we could perhaps pursue the analysis to find the average path length of ILBT's. But really, Mr. Quick, your algorithm still does not deserve to be implemented. The average path length will be somewhere between $2n \ln n$ and $(1/\ln 2)\, n \ln n$; the extra time your method takes at each node slows the program down so much that the slightly smaller path length is pointless. It was clear from the start that ILBT's would lose out to other methods in all respects (space, time, ease of implementation, and so on). The only saving feature was that your algorithms lead to instructive mathematics that we might someday be able to apply to the analysis of a really useful method. You undoubtedly knew that too, so thanks for the ideas."

Index

Progress in Computer Science and Applied Logic

Progress in Computer Science and Applied Logic is a series that focuses on scientific work of interest to both logicians and computer scientists. Thus both applications of mathematical logic will be topics of interest. An additional area of interest is the foundations of computer science.

The series (previously known as *Progress in Computer Science*) publishes research monographs, graduate texts, polished lectures from seminars and lecture series, and proceedings of focused conferences in the above fields of interest. We encourage preparation of manuscripts in such forms as LaTEX or AMS TEX for delivery in camera-ready copy, which leads to rapid publication, or in electronic form for interfacing with laser printers or typesetters.

Proposals should be sent directly to the editors or to:
Birkhäuser Boston, 675 Massachusetts Ave., Suite 601, Cambridge, MA 02139

Progress in Computer Science and Applied Logic

PCS 1 Mathematics for the Analysis of Algorithms, 3rd Edition
Daniel H. Greene & Donald E. Knuth

PCS 2 Applied Probability – Computer Science: The Interface, Volume I
Edited by Ralph L. Disney & Teunis J. Ott

PCS 3 Applied Probability – Computer Science: The Interface, Volume II
Edited by Ralph L. Disney & Teunis J. Ott

PCS 4 Notes on Introductory Combinatorics
George Pólya, Robert E. Tarjan, & Donald R. Woods

PCS 5 The Evolution of Programs
Nachum Dershowitz

PCS 6 Lecture Notes on Bucket Algorithms
Luc Devroye

PCS 7 Real-Time Control of Walking
Marc D. Donner

PCS 8 Logic for Computer Scientists
Uwe Schöning

PCS 9 Feasible Mathematics
Sam Buss & Phil Scott